# A&P TECHNICIAN
# AIRFRAME
# WORKBOOK

JEPPESEN®
Sanderson Training Products

JS322701C

# Preface

Certified Aviation Maintenance Technician Schools all operate under individual Federal Aviation Administration approvals. Within each school's approved curriculum there is enough information to meet the FAA knowledge requirements for certification as a technician.

Realizing the need for updated information, presented in a format that keeps with the latest developments in education technology, Jeppesen is offering an Integrated Training Program (ITP). This program presents the basic information needed for technician certification in a form that provides a good foundation in aviation maintenance knowledge. The ITP training manuals, workbooks, study guides, and audiovisuals are designed to follow the FAA written examinations and are written to the level required by FAR Part 147 schools.

All of these materials can be used by an individual preparing for certification on their own, or by schools preparing groups of individuals for certification under FAR Part 65.

One of the most basic premises of aviation maintenance is that all operations must be carried out according to FAA-approved data. Furthermore, all information of a specific nature must be furnished by manufacturers of aircraft, powerplants, or components and be followed in detail. These training manuals are, of necessity, general in nature and cover a wide scope of aircraft, powerplants, and components. Because of this, it must be understood that these training materials should *never* take precedence over specific information furnished by a manufacturer.

The authors and publishers of this Integrated Training Program wish to express our appreciation to the various manufacturers, Aviation Maintenance Technician Schools, and FAA personnel who have aided our efforts to provide this coordinated training material to meet the challenge of better training for aviation maintenance technicians.

# Table of Contents

# Chapter I
# Aircraft Structures

1.  Airplanes and helicopters are supported by _____ lift.

2.  The angle between the chord of the wing and the relative wind is known as the angle of _____.

3.  Air passing across an airfoil will create a region of _____ (low or high) pressure above the airfoil.

4.  The center of lift of an aircraft is usually located _____ (ahead or behind) the center of gravity.

5.  The main spanwise structural member of a wing is the _____.

6.  The top of a wing spar is subjected to _____ stresses in flight.

7.  The bottom of a wing spar is subjected to _____ stresses in flight.

8.  Fabric-covered airplane wings have a(n) _____-type structure.

9.  What component attaches to the spars to give the wing the aerodynamic shape it needs to produce lift?

    _____

10. The steel wire that runs from the front spar inboard to the rear spar outboard in a truss-type wing is called a(n) _____ wire.

11. A wing which uses no external struts or braces to support it is called a(n) _____ wing.

12. Milled skins for high-speed aircraft may be produced by conventional machining or by _____ milling, or by _____ machining.

13. Laminated structural material such as bonded honeycomb is used for aircraft structure because it provides _____ as well as a favorable strength/weight ratio.

14. A fuel tank that uses a sealed portion of the aircraft structure to hold fuel is called a(n) _____ tank.

15. The three axes of an airplane are:
    a. _____ or _____
    b. _____ or _____
    c. _____ or _____

16. The elevators rotate an airplane about its _____ axis.

17. The tail load of an airplane in level flight normally acts _____ (upward or downward).

18. When the control wheel of the aircraft is pulled back, the trailing edge of the elevator moves _____ (up or down).

19. The ailerons rotate an airplane about its _____ axis.

20. To bank an airplane to the right, the aileron on the right wing moves _____ (up or down).

21. Which aileron travels a greater distance, the one moving up or the one moving down? The one moving _____ (up or down).

22. A(n) _____ -type aileron has its hinge line located far enough back that its leading edge will protrude below the wing surface when the aileron is raised.

23. The rudder will rotate an aircraft about its _____ axis.

24. The temporary movement of the nose of an airplane toward the wing that is rising at the beginning of a turn is called _____ yaw.

25. The _____ is used to overcome the effects of adverse yaw.

26. Depressing the right rudder pedal will move the trailing edge of the rudder to the _____ (right or left).

27. Control surfaces are balanced so that their center of gravity is _____ (ahead of or behind) the hinge line.

28. Stiffness of the thin sheet metal covering for a control surface may be increased by _____ the metal skin.

29. On jet transport aircraft equipped with two sets of ailerons, only the _____ (inboard or outboard) is used during high speed flight.

30. A(n) _____ is a control device that destroys lift by disrupting the airflow over a part of the wing.

31. The assembly of tail surfaces of an airplane is called the _____.

32. The fixed vertical tail surface of an airplane is used to provide stability about the _____ axis.

33. The movable vertical tail surface is called the _____.

34. The fixed horizontal tail surface is called the horizontal _____.

35. The movable horizontal tail surfaces are called the _____.

36. The extension of the vertical stabilizer that may extend nearly to the cabin section is known as a(n) _____ fin.

37. An all movable horizontal tail surface is called a(n) _____.

38. An all movable tail surface usually has a large tab installed on its trailing edge. This is known as a(n) _____ (servo or anti-servo) tab.

39. The movable surfaces on a V-tail are known as _____.

40. When wing flaps are lowered, they increase both the lift and the _____ that is produced by the wing.

41. When plain flaps are lowered, they _____ (increase or decrease) the camber of the wing.

42. _____ (what type) flaps are used to prevent the airflow from breaking away from their upper surfaces when the flaps are fully extended.

43. A(n) _____ (what type) flap rides out of the trailing edge of the wing on tracks and increases the wing area as well as its camber.

44. A drooped leading edge flap increases the _____ of the wing.

45. A(n) _____ (what type) flap is a special form of leading edge flap that gives the leading edge a high lift shape.

46. A fixed slot in an airplane wing is usually located ahead of the _____ (aileron or flap).

47. Leading edge slats may be extended by one of two methods:
    a. _____
    b. _____

48. A stall strip is a small triangular strip of metal installed on the leading edge of a wing _____ (at the wing root or ahead of the aileron).

49. _____ (what unit) are pairs of small, low aspect ratio airfoil sections mounted on the upper surface of a wing to pull high-energy air down onto the surface to prevent shock-induced separation.

50. The three primary controls of an airplane are:

    a. _____

    b. _____

    c. _____

51. A(n) _____ trim tab is usually mounted on a control surface to correct for an out-of-trim condition.

52. An adjustable trim tab on the trailing edge of the elevator should be moved _____ (up or down) to trim the airplane nose up.

53. A balance tab moves in the _____ (same or opposite) direction as the control surface to which it is attached.

54. A servo tab moves in the _____ (same or opposite) direction as the control surface to which it is attached.

55. An anti-servo tab moves in the _____ (same or opposite) direction as the control surface to which it is attached.

56. A spring tab moves only when the control forces are _____ (high or low).

57. An aerodynamic balance panel gives the most assistance to the movement of the control surface during _____ (large or small) deflections of the surface.

58. In a _____ (Pratt or Warren) truss fuselage, the stays carry only the tensile loads.

59. Both tensile and compressive loads are carried in the diagonal members of a _____ (Pratt or Warren) truss fuselage.

60. A(n) _____ (what type) structure carries all of the stresses in its skin.

61. A(n) _____ (what type) fuselage has a substructure to stiffen the external skin.

62. A structure that is built with more than one path for the stresses so a crack will not destroy the structure is called a(n) _____ design.

63. Aircraft using a tailwheel type landing gear configuration are also called _____ gear airplanes.

64. Aircraft using a nosewheel type landing gear configuration are also called _____ gear airplanes.

65. When a landing gear is retracted into the structure, the _____ (parasite or induced) drag is reduced.

66. Today, almost all piston-powered airplanes enclose the engine in a(n) _____ cowling.

67. Heat is removed from the cylinders of an air-cooled engine by forcing air to flow through _____ on the cylinders.

68. The amount of airflow through the fins of a high-powered engine is usually controlled by _____ at the air exit.

69. Cowl flaps are normally _____ (open or closed) during ground operations.

70. The two most common locations for turbojet or turbofan engine installations are:

   a. _____

   b. _____

71. Helicopters utilize a great deal of _____ material fabrication, both in the fuselage and the rotor blades.

# Chapter II
# Assembly And Rigging

## PART ONE

1.  The air surrounding the earth is composed of roughly _____ percent nitrogen and _____ percent oxygen, and about one percent of other gases.

2.  State the standard sea level pressure in each of the following units:

    a.  pounds per square inch        _____

    b.  inches of mercury             _____

    c.  millibars                     _____

3.  What instrument in the cockpit of aircraft measures the absolute pressure of the atmosphere surrounding the craft?

    _____

4.  The four temperature scales used in aeronautical computations are:

    a.  _____

    b.  _____

    c.  _____

    d.  _____

5.  Convert the following temperatures:

    a.  100 degrees Fahrenheit   =   _____ degrees Celsius

    b.  217 degrees Celsius      =   _____ degrees Fahrenheit

    c.  5,280 degrees Kelvin     =   _____ degrees Celsius

    d.  700 degrees Rankine      =   _____ degrees Fahrenheit

6. The standard sea level temperature for aeronautical computations are:

    a. _____ degrees Celsius

    b. _____ degrees Fahrenheit

7. _____ is the ratio between the amount of moisture present in a given amount of air at a specific temperature and pressure, and the maximum amount it could hold under the same circumstances.

8. Humid air is _____ (less or more) dense than dry air.

9. Air density is expressed in terms of _____ (what unit) per cubic foot.

10. Standard sea level air density is _____ per cubic foot.

11. _____ altitude is the altitude in standard air that has the same density as the ambient air.

12. The two forms of energy found in air are:

    a. _____

    b. _____

13. Air pressure is considered to be a form of _____ energy.

14. Air velocity is considered to be a form of _____ energy.

15. Bernoulli's principle states that if the total amount of energy in the air remains constant, any increase in kinetic energy will cause an equal _____ (increase or decrease) in potential energy.

16. Subsonic aerodynamics considers the airflow over a wing to be _____ (compressible or noncompressible).

17. The maximum thickness of a subsonic airfoil occurs about _____ of the way back from the leading edge.

18. The point on the leading edge of an airfoil where the airflow separates, some going over the top and some below is called the _____ point.

19. The amount of lift produced by aerodynamic action is determined by these three factors:

    a. _____

    b. _____

    c. _____

20. Two variables that affect the coefficient of lift of an airfoil are:

    a. _____

    b. _____

21. When calculating lift, the symbol used for the dynamic pressure of the air moving over the airfoil is _____.

22. The formula for calculating the dynamic pressure of air is:

    _____ = _____

23. The random-flowing layer of air on the surface of a lift-producing airfoil is called the _____ layer.

24. Two types of drag that act on an aircraft in flight are:

    a. _____

    b. _____

25. Induced drag _____ (increases or decreases) as the airspeed increases.

26. Parasite drag _____ (increases or decreases) as the airspeed increases.

27. The center of pressure remains relatively stationary on a(n) _____ (symmetrical or asymmetrical) airfoil.

28. The three axes of an airplane are:

    a. _____

    b. _____

    c. _____

29. The two types of stability exhibited by an aircraft are:

    a. _____

    b. _____

30. About which of the three axes of an airplane do each of these items provide stability:

    a. Horizontal tail surfaces: _____

    b. Dihedral in the wing: _____

    c. Vertical fin on the tail: _____

31. The corkscrewing slipstream from the propeller requires the pilot to use _____ (right or left) rudder on takeoff.

32. About which axis of an airplane does each of these control surfaces cause the airplane to rotate?

    a. Aileron: _____

    b. Elevators: _____

    c. Rudder: _____

33. An elevator downspring is used to provide a mechanical force on the elevators as a safety factor when the airplane is flown with its center of gravity near its _____ (forward or rearward) limit.

34. The aileron moving _____ (up or down) travels a greater distance from its neutral position when the control wheel is moved to its limit.

35. The cable that connects the two ailerons together and does not attach to the control wheel is called the _____ cable.

36. An aileron with its hinge point back a way from its leading edge is called a(n) _____-type aileron.

37. Many modern airplanes have their aileron and rudder controls interconnected with _____ (springs or steel cables).

38. A balance tab moves in the _____ (same or opposite) direction as the surface to which it is attached.

39. Stabilators are normally equipped with _____ (servo or anti-servo) tabs.

40. An anti-servo tab moves in the _____ (same or opposite) direction as the surface on which it is attached.

41. Moving the leading edge of an adjustable stabilizer up, trims the airplane nose-_____ (up or down).

42. Wing flaps that move out of the trailing edge of the wing on tracks to increase the area as well as the camber are _____ (what type) flaps.

43. A fixed _____ (slot or slat) ducts air over the top of the wing to keep the aileron effective and provide lateral control during most of the stall.

44. Ideally, a wing should stall at the _____ (root or tip) first.

45. The _____ is a simple method to stop, or reduce, the spanwise flow of air on a wing.

46. Two types of devices used by the Boeing 727 for lateral control are:

    a. _____

    b. _____

47. What devices are used to move the elevators on a Boeing 727 in the event of the failure of both of the hydraulic systems?

    _____

48. The Boeing 727 has _____ (how many) independent rudders to provide yaw control.

49. Dutch roll is prevented in the Boeing 727 by the use of a(n) _____ .

50. What two devices, other than the triple-slotted Fowler flaps, alter the camber of a Boeing 727 when the flaps are lowered?

    a. _____

    b. _____

51. High-speed aerodynamics deal with air as though it were a(n) _____ (compressible or incompressible) fluid.

52. The speed of sound at standard sea level conditions is _____ knots.

53. The ratio of the true airspeed of an aircraft to the speed of sound is known as the

    _____  _____.

54. Give the names of the realms of flight associated with each of these speed ranges:

    a.  Below Mach 0.75:                    _____

    b.  Between Mach 0.75 and Mach 1.20:    _____

    c.  Above Mach 1.20:                    _____

55. When air flows through an oblique shock wave, its velocity _____ (increases or decreases).

56. When air flows through a normal shock wave, its velocity _____ (increases or decreases).

57. When air flows through an expansion wave, its static pressure _____ (increases or decreases).

58. The aerodynamic center of an airfoil in supersonic flight is located approximately _____ (what percent) of the way back from the leading edge.

59. The first shock wave to form on a subsonic airfoil in transonic flight forms on the _____ (top or bottom) of the wing.

60. The flight Mach number at which there is the first indication of sonic flow is called the _____ of the airfoil.

61. To assure the air entering the compressor of a turbine engine is slowed to subsonic speed, a(n) _____ (normal or oblique) shock wave may be formed across the engine inlet duct.

62. One of the biggest problems of hypersonic flight is _____ _____ of the aircraft structure.

## PART TWO

1. The rotor of an autogiro is turned by _____ forces.

2. Two ways of counteracting torque from a helicopter rotor system are:

   a. _____

   b. _____

3. Rigid rotor systems have movement only about their _____ (flap, drag, or feather) axis.

4. The four forces acting on a helicopter rotor are:

   a. _____

   b. _____

   c. _____

   d. _____

5. Most helicopters use a(n) _____ (symmetrical or asymmetrical) airfoil section.

6. What are the two gyroscopic forces that act on the spinning rotor of a helicopter?

   a. _____

   b. _____

7. _____ (Precession or Coriolis effect) causes the rotor to tilt 90 degrees to the point the control force is applied.

8. _____ (Precession or Coriolis effect) causes a rotor blade to attempt to speed up as it flaps up and moves its center of mass closer to the rotor mast.

9. Coriolis effect is minimized on a fully articulated rotor by the use of the _____ (drag, flap, or feather) hinge.

10. The coriolis effect is minimized on a semi-rigid rotor by using an _____ (underslung or offset) hub.

11. Torque of the main rotor is compensated on a single-rotor helicopter by the use of a(n) _____.

12. What is meant by the abbreviation IGE?

    _____

13. What is meant by the abbreviation OGE?

    _____

14. Ground effect is normally considered to extend upward from the surface about _____ (what part) of a rotor span.

15. A helicopter experiences dissymmetry of lift in what flight condition? _____ (hovering or forward flight).

16. Dissymmetry of lift is compensated for by blade _____.

17. The helicopter blade that moves in the same direction that the helicopter is moving is called the _____ (advancing or retreating) blade.

18. A helicopter rotor normally stalls on the _____ (advancing or retreating) blade.

19. A helicopter rotor stalls in _____ (high- or low-) speed flight.

20. Lift produced by a helicopter rotor system because of forward flight is called _____ lift.

21. If a helicopter experiences engine failure, the rotor is turned by an aerodynamic force known as the _____ force.

22. During powered flight airflow through the rotor is _____ (upward or downward).

23. During autorotation, the airflow through the rotor of a helicopter is _____ (upward or downward).

24. The helicopter control that changes the pitch on all the rotor blades at the same time is called the _____ pitch control.

25. The helicopter control that changes the pitch of the rotor blades at a certain point in their rotation is called the _____ pitch control.

26. The fuselage of some helicopters is kept relatively level in flight at high airspeed by using a(n) _____ controlled by the cylic pitch control to increase the down load on the tail as the cyclic control is moved forward.

27. Boosted control on a helicopter cannot feed vibration back through the controls if an _____ control system is used.

28. The pitch of the tail rotor is controlled by _____ operated by the pilot.

29. A helicopter is inherently _____ (stable or unstable).

30. Three systems used to increase the stability of a helicopter are:

    a. _____

    b. _____

    c. _____

31. Low frequency vibration is felt as a(n) _____ (beat or buzz).

32. Vibrations caused by the main rotor system are usually of the _____ (low or high) frequency type.

33. An out-of-balance rotor would normally cause a _____ (lateral or vertical) vibration.

34. An out-of-track rotor would normally cause a _____ (lateral or vertical) vibration.

35. Helicopter blades must be balanced both:

    a. _____

    b. _____

36. Three methods of checking main rotor blade track are:

    a. _____

    b. _____

    c. _____

37. Adjusting the length of the pitch change rods is normally done to correct an out-of-track condition found _____ (on the ground or in flight).

38. Bending the rotor blade trim tabs is normally done to correct an out-of-track condition found _____ (on the ground or in flight).

39. Cooling air is supplied to a reciprocating engine installed in a helicopter by a(n) _____.

40. Helicopter power output is controlled by the _____ pitch control.

41. Two types of turboshaft engines used to power a helicopter are:

    a. _____

    b. _____

42. Helicopters powered by reciprocating or direct-shaft turbine engines must use some form of _____ to disconnect the engine from the transmission during starting.

43. Helicopter must have some sort of _____ device to release the engine from the rotor any time the speed of the engine drops below that required to drive the rotor.

## PART THREE

1. The angle between the chord line of the wing and the longitudinal axis of the airplane is called the angle of _____.

2. When an airplane's wing is washed in, its angle of incidence is _____ (increased or decreased).

3. Control surface travel specifications for a particular make and model of aircraft may be found in what FAA publication?

_____

4. Some aircraft aileron system are rigged so that when there is no airload both ailerons will be a few degrees below the trailing edge of the wing. This is known as aileron _____.

5. Fill in the strand number designation for each of these cable types:
   a. Non-flexible          _____ or _____
   b. Flexible              _____
   c. Extra-flexible        _____

6. What percentage of the full cable strength may be obtained with each of these terminations?
   a. Woven splice          _____%
   b. Nicopress sleeve      _____%
   c. Swaged terminal       _____%

7. After installing the terminals on a control cable, it should be subjected to a proof load test which is _____ (what percent) of the breaking strength of the cable.

8. Aircraft control cables should be carefully checked for what two defects?

   a. _____

   b. _____

9. Fairleads _____ (may or may not) be used to change the direction of a cable.

10. Using the cable tension chart (figure 2-136), on page 102 of the text, find the correct tension for the following conditions:

    a. 3/16-inch 7 × 19 cable when adjusted at 40 degrees Fahrenheit =

       _____ pounds

    b. 1/8-inch 7 × 19 cable when adjusted at 40 degrees Fahrenheit =

       _____ pounds

    c. 1/8-inch 7 × 19 cable when adjusted at 90 degrees Fahrenheit =

       _____ pounds

    d. 1/16-inch 7 × 7 cable when adjusted at 70 degrees Fahrenheit =

       _____ pounds

11. For a turnbuckle to develop its full strength, there must be no more than _____ (how many) threads exposed at either end of the barrel.

12. When safety-wiring a turnbuckle, you should have at least _____ (how many) full wraps of wire around the shank before the wire is cut off.

13. When installing threaded rod end fittings for push-pull type controls, if you can pass a piece of safety wire through the check hole, the rod end _____ (is or is not) screwed in far enough.

14. When the upper wing of a biplane is ahead of the lower wing, the airplane is said to have _____ (positive or negative) stagger.

15. The difference between the angle of incidence between the two wings of a biplane is known as _____.

16. The streamlined wires on a biplane that run from the center section to the interplane strut on the lower wing are called the _____ (flying or landing) wires.

17. When assembling a biplane, the _____ (upper or lower) wing panels are installed first.

## PART FOUR

1. The Type Certificate issued to an aircraft is an approval for its _____ (design or production).

2. Airworthiness Directives _____ (are or are not) sent to every registered owner of the type of aircraft affected.

3. Federal Aviation Regulation Part _____ (what number) specifies the inspections that are required to determine the condition of airworthiness of an aircraft.

4. If an aircraft being operated under FAR 91, Subpart E undergoes an annual inspection on March 18, 1991, the inspection will remain valid until
_____.

5. An Inspection Authorization (IA) _____ (is or is not) required of the mechanic conducting an annual inspection.

6. The scope and detail of the annual inspection may be found where in the Federal Aviation Regulations?
_____

7. A list of discrepancies that prevent an airplane passing a 100-hour inspection _____ (is or is not) required to be sent to the owner of the airplane and to the FAA.

8. Aircraft operating under Instrument Flight Rules must have their altimeters and static systems checked each _____ months.

9. Radar beacon transponders must be checked every _____ months.

10. Large aircraft and multi-engine turbine aircraft must be maintained in accordance with FAR Part 91, Subpart _____.

11. If an aircraft is being operated for hire it must be inspected every _____ (what interval), in addition to the annual inspection.

12. The scope and detail of the annual and the 100-hour inspection _____ (are or are not) the same.

13. Compliance with aircraft manufacturer's Service Bulletins _____ (are or are not) mandatory.

# Chapter III
# Aircraft Fabric Covering

1.  The clear dope film and organic fabric such as cotton and linen are protected from damage by the sun by mixing extremely fine flakes of _____ in with the dope.

2.  Which is more flammable, cellulose nitrate or cellulose butyrate?
    Cellulose _____

3.  Recovering an aircraft in the same manner as was used by the manufacturer _____ (is or is not) considered to be a major repair.

4.  Three sources of approved data for recovering an airplane are:
    a. _____
    b. _____
    c. _____

5.  Three categories of fabric covering are:
    a. _____
    b. _____
    c. _____

6.  The minimum tensile strength of new grade-A cotton fabric is _____ pounds per square inch.

7.  The fabric on an aircraft with a wing loading of 9.5 pounds per square foot and a never-exceed speed of 180 miles per hour can deteriorate to _____ pounds per square inch before is considered to be unairworthy.

8.  _____ fabric may be used to cover a plywood structure to give it an extremely smooth finish.

9.  _____ tape is used for inter-rib bracing to hold the wing ribs in their upright position until the fabric is stitched to them.

10. _____ tape is used to cover all of the seams, over all of the ribs, around the tips, and along the trailing edge of all the surfaces.

11. Drainage grommets are usually applied with the _____ (what number) coat of dope.

12. _____ is a special slow-drying solvent that prevents rapid evaporation of aircraft dope.

13. The dope in which the fungicidal paste is mixed should be applied to the fabric _____ (thinned or full strength).

14. One pound of aluminum powder should be mixed with _____ gallons of unthinned clear dope, for the aluminum dope coats.

15. Rejuvenator is a mixture of potent solvents and _____.

16. Rejuvenator _____ (will or will not) restore strength to deteriorated fabric.

17. All of the interior wood members of an aircraft should be given a coat of _____ before the structure is recovered.

18. Two methods of applying the fabric to an aircraft structure are:
    a.  The _____ method.
    b.  The _____ method.

19. A(n) _____ stitch is used for hand sewing the fabric along the trailing edge of the wing.

20. The wrinkles are removed from cotton fabric after it is installed on an aircraft structure by spraying it with _____.

21. The first coat of dope applied to cotton fabric will cause the fabric to _____ (tighten or loosen).

22. The brush recommended for applying aircraft dope should have _____ (animal or synthetic) bristles.

23. Anti-tear strips should be used under the reinforcing tape on aircraft having a never-exceed speed in excess of _____ miles per hour.

24. A wing rib of an airplane with a never-exceed speed of 200 miles per hour should have a stitch spacing ever _____ inches in the slipstream and every _____ inches outside of the slipstream.

25. Rib stitch knots should be placed _____ (in the center or beside) the reinforcing tape on a wing rib.

26. A modified _____ knot is used to secure rib stitches.

27. A(n) _____ knot is used to join two pieces of waxed rib stitch cord.

28. Before a fabric-covered aircraft structure is dry-sanded, it should be electrically _____.

29. The surface tape along the trailing edge of a control surface should be notched every 18 inches if the airplane has a never exceed speed of _____ miles per hour or more.

30. Drainage grommets should be opened by _____ (cutting or punching) the center out.

31. Separation of the finish can be caused by too _____ (much or little) aluminum powder in the dope used for the aluminum dope coats.

32. The glass-like surface of a doped finish comes from the _____ (clear or aluminum) dope coats.

33. Properly applied aluminum dope coats will dry with a slightly _____ (glossy or dull) finish.

34. Dacron® is a(n) _____ (polyester, polyamide, or acrylic) fabric.

35. Two weights of Dacron fabric used for aircraft covering are:

    a. _____ ounces per square yard.

    b. _____ ounces per square yard.

36. _____ is used to shrink Dacron fabric on an aircraft structure.

37. _____ dope should be used as the fill coats for a doped finish on Dacron fabric.

38. _____ (Butyrate or Nitrate) dope should be used on treated glass cloth that is used to cover an airframe.

39. The first coat of dope applied to a glass cloth covering should be _____ (brushed or sprayed) on.

40. When making a sewed repair to an L-shaped tear in an aircraft covering, you should use a(n) _____ stitch and should start sewing at the _____ (apex or end) of the tear.

41. When using a baseball stitch to repair an L-shaped tear, you should use a minimum of _____ stitches per inch, and lock the stitches every _____ to _____ stitches with a modified seine knot.

42. When repairing the fabric of an aircraft whose never-exceed speed is greater than 150 miles per hour, you should _____ (sew or dope) the patch in place.

# Chapter IV
# Aircraft Painting
# And Finishing

1. _____ are used in aircraft dope to make its film flexible.

2. Rejuvenators restore the _____ to a dried-out dope film.

3. A dope film should be scrubbed with toluol or _____ before the rejuvenator is sprayed on.

4. Rejuvenator is a mixture of potent _____ and _____.

5. The finish coats of dope should be sprayed on with a _____ (heavy or light) coat.

6. The colored dope may be given a glossy finish by mixing it with up to _____% clear dope and some retarder.

7. The topcoats of a doped finish may peel off if there is _____ (too much or too little) aluminum powder in the aluminum dope coats.

8. A dope film blushes when _____ condenses out of the air into the freshly applied dope film.

9. _____ mixed with the dope will prevent blushing if the conditions are not too severe.

10. The relative humidity of the air may be lowered to prevent blushing by _____ (warming or cooling) the air.

11. Two causes for pinholes in a dope film are:

    a. _____

    b. _____


12. Three causes for runs and sags in a dope surface are:

    a. _____

    b. _____

    c. _____


13. Orange peel in a dope surface may be caused by:

    a. _____

    b. _____


14. Localized spots within the dope film that do not dry are called _____.


15. Dope that is brushed on too heavy or brushed when it is too cold will cause dope
    _____.


16. Two disadvantages of a polyurethane finish on a fabric-covered surface are:

    a. _____

    b. _____


17. A(n) _____ is added to the newer polyurethane materials to
    make them suitable for use over fabric surfaces.


18. Polyurethane material should be removed from a damaged surface with
    _____ before a repair can be made.


19. Which type of finish will paint stripper penetrate and swell up, causing it to break
    its bond with the primer, enamel or lacquer? _____


20. Parts of an aircraft surface that must not be touched with paint stripper may be
    masked with:

    a. _____

    b. _____

21. Paint stripper should be brushed on in a _____ (thick or thin) layer.

22. Stripper residue is cleaned from an aircraft surface with _____ or
    _____.

23. New media blasting systems use _____ material to remove the
    old finish and surface corrosion.

24. Filiform corrosion may occur under the dense film of _____ (polyure-
    thane or acrylic) topcoats.

25. A(n) _____ coating changes the surface of an aluminum alloy
    into an oxide film that is chemically inert and will not allow filiform corrosion.

26. Most high-volume production all-metal aircraft are finished with an
    _____ lacquer.

27. Acrylic lacquer has a _____ (high or low) solids content.

28. The best finish produced by an acrylic lacquer will result from _____ (a or b).
    a.  A few heavy coats of finish sprayed on
    b.  Several light coats of finish sprayed on

29. The most durable finish system and the one that gives the most pleasing appear-
    ance is produced by the _____ system.

30. _____ primer may be used under a polyurethane topcoat, but if
    it is not properly cured, it may cause filiform corrosion.

31. After mixing polyurethane with its catalyst, it must be allowed to stand for 15 to 30
    minutes. This is called the _____ time of the material.

32. The viscosity of a thinned finishing material is determined using a(n)
    _____.

33. A polyurethane finish is usually dry enough to be taped after _____ hours of
    drying time, but it is best to wait 24 hours.

34. The time between the mixing of a catalyzed material and the time it has set up too much for it to be used is called the _____ life of the material.

35. Wash primer is made up of these three components:

    a. _____

    b. _____

    c. _____

36. Wash primers should be applied with a maximum film thickness of _____ mil.

37. If the topcoat is not applied over the wash primer within _____ hours, another coat of primer must be applied over the first one.

38. About the most critical aspect of the application of wash primers is the necessity of having sufficient _____ in the air to properly convert the acid into a phosphate film.

39. Indicate below the three layers necessary when applying high-visibility finishes:

    a. _____

    b. _____

    c. _____

40. An acid-proof finish, far superior to asphaltum paint, is a good coat of _____ enamel.

41. Asphaltum products such as those used on battery boxes and float bottoms may be thinned using _____.

42. Zinc chromate paste used for making leakproof seams _____ (will or will not) harden.

43. _____ oil is used to protect the inside of the tubular structure in aircraft fuselages, empennage structure, and landing gear.

44. Most of the fumes from finishing materials are _____ (heavier or lighter) than air.

45. All filter units and water traps in the shop air system should be drained _____ (what interval).

46. The three systems used to spray liquid finishing materials are:

    a. _____

    b. _____

    c. _____

47. Filter-type masks when worn by spray painters _____ (will or will not) filter out the paint fumes.

48. Paint mixing agitators should be driven by a(n) _____ drill motor, never with an _____ drill motor.

49. A correctly adjusted spray gun should produce a uniform fan-shaped spray, with the fan _____ (parallel or perpendicular) to the wing ports.

50. Match the spray pattern with the number that identifies the cause of the distortion.

    a. _____

    b. _____

    c. _____

    d. _____

1. Plugged-up wing port hole on one side
2. Material too thick for the amount of atomizing air.
3. Air leaking into the fluid line.
4. Too much air through the wing port holes for the amount of material that is being sprayed.
5. Material buildup in the air cap.
6. Material in the cup is too thin.

51. An excessive overspray may be caused by too _____ (much or little) atomizing air pressure.

52. "When painting an airplane, you should paint the corners and edges first, then the flat surfaces." This statement is _____ (true or false).

53. Each pass made by a paint spray gun should overlap the previous pass by about _____ (what fraction) of their width.

54. When an airplane is sprayed, the painting should be done so the overspray falls _____ (ahead of or behind) the area being painted.

55. Hardened dope and lacquer may be removed from a spray gun by soaking it in _____ or _____.

56. The location and size of registration numbers required on an aircraft are specified in Federal Aviation Regulations Part _____.

57. The Roman letter _____ preceding the registration numbers must be placed on all aircraft registered in the United States of America.

58. How much length would be required to lay out the numbers N469AB, if the numbers are 12 inches high?
    _____ inches.

59. A ball-point pen is a _____ (good or bad) tool to use to mark the surface of an aircraft when laying out the registration numbers.

60. Decals are applied by soaking them in clean warm _____ until the clear portion slides easily from the backing.

61. Dried overspray on the paint shop floor should be removed by _____ (wet or dry) sweeping it.

# Chapter V
# Aircraft Metal
# Structure Repair

1.  The type of metal most often used in the construction of civilian aircraft is heat-treated _____ alloy.

2.  The two basic types of sheet metal structure used for aircraft are:

    a.  _____

    b.  _____

3.  A repair to an aircraft structure must restore its original:

    a.  _____

    b.  _____

    c.  _____

4.  The five basic stresses encountered by an aircraft structure are:

    a.  _____

    b.  _____

    c.  _____

    d.  _____

    e.  _____

5.  The top of a wing spar, in flight, is subjected to a _____ (tensile or compressive) stress.

6.  The bottom of a wing spar, in flight, is subjected to a _____ (tensile or compressive) stress.

7.  A torsional stress is made up of _____ and _____ acting perpendicular to each other.

8.  A properly designed rivet joint is subjected what type of stress?

    _____

9.  To prevent a small crack from extending, you may _____ its end.

10. Aluminum alloys _____ (are or are not) susceptible to corrosion.

11. Name the primary alloying element in each of these aluminum alloys:
    a.  2117        _____
    b.  2024        _____
    c.  5056        _____
    d.  7075        _____

12. Commercially pure aluminum is identified by the number _____.

13. Most of the aluminum alloy used for aircraft structure is alloy number
    _____.

14. When an aluminum alloy is heated to its critical temperature and quenched, it is
    _____ (solution or precipitation) heat treated.

15. Artificial aging is another name for _____ (solution or precipitation)
    heat-treatment.

16. Aluminum alloys may be _____ (hardened or softened) by annealing.

17. Give the temper designation for each of these conditions:
    a.  As fabricated                               _____
    b.  Annealed                                    _____
    c.  Solution heat treated                       _____
    d.  Solution heat treated and cold worked       _____
    e.  Solution heat treated and artificially aged _____
    f.  Strain hardened to quarter-hard temper      _____
    g.  Strain hardened to full-hard temper         _____

18. Three methods of corrosion prevention used on aluminum alloys are:

    a. _____

    b. _____

    c. _____

19. Three requirements for the formation of corrosion on aluminum alloy are:

    a. _____

    b. _____

    c. _____

20. Pure aluminum _____ (will or will not) corrode.

21. A protective oxide film may be formed on the surface of a piece of aluminum alloy by what two methods?

    a. _____

    b. _____

22. Aluminum alloy-faced honeycomb panels are useful as a structural material because of their _____ as well as strength.

23. Magnesium is _____ (lighter or heavier) than aluminum.

24. Three drawbacks to the use of magnesium as a structural material are:

    a. _____

    b. _____

    c. _____

25. When using a scale for sheet metal layout, you _____ (should or should not) measure from its end.

26. Felt marking pens _____ (are or are not) widely used for sheet metal layout.

27. What type of punch is used to locate the center of a rivet hole when using the old skin as a pattern for a new one?

    A(n) _____ punch.


28. Identify the color code for the handles of the aviation snips which:

    a.  Cut right        _____

    b.  Cut left          _____

    c.  Cut straight    _____


29. The three main parts of a twist drill are:

    a.  _____

    b.  _____

    c.  _____


30. A number _____ drill is used to open holes for 1/8" diameter rivets.


31. The _____ brake is the most generally used bending machine in aircraft maintenance shops.


32. What color is used to identify each of these Cleco fasteners?

    a.  3/32-inch _____

    b.  1/8-inch _____

    c.  5/32-inch _____

    d.  3/16-inch _____


33. A(n) _____ is used for locating new rivet holes in undrilled skins.


34. Identify the head shape of each of these solid aluminum alloy rivets:

    a.  AN426 (MS20426)      _____

    b.  AN470 (MS20470)      _____

    c.  AN430                  _____

    d.  AN442                  _____

35. Identify the alloy used in solid aluminum rivets having these markings on their heads:

    a.      no mark     _____

    b.      dimple     _____

    c.      teat     _____

    d.      raised bars     _____

    e.      raised cross     _____

36. Give the alloy number used for rivets identified by the following code letters:

    a.   A    _____

    b.   B    _____

    c.   AD   _____

    d.   D    _____

    e.   DD   _____

    f.   E    _____

37. Magnesium skins should be riveted using a rivet identified by the code letter
_____.

38. Rivet length is measured in increments of _____-inch.

39. Rivet diameters are measured in increments of _____-inch.

40. Which measurement (L) is correct for the length of this rivet?

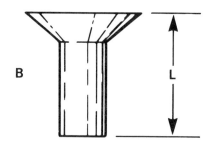

41. The single-shear strength of an AD3 rivet is _____ pounds.

42. The double-shear strength of a DD4 rivet is _____ pounds.

43. What is the bearing strength of a sheet of 0.032-inch 2024-T3 clad aluminum alloy for the following rivet sizes?

    a.   3/32-inch _____ pounds

    b.   1/8-inch   _____ pounds

    c.   5/32-inch _____ pounds

44. For a rivet joint to develop its required strength, no rivet should be installed with its center nearer the edge of a sheet than _____ rivet diameters.

45. Adjacent rivets in a row should be no closer than _____ diameters, and no further apart than _____ to _____ diameters of the rivet shank.

46. The distance between rows of rivets should be _____ (what fraction) of the distance between the rows.

47. What size twist drill should be used for the installation of each of these rivet sizes?

    a.   3/32-inch _____

    b.   1/8-inch   _____

    c.   5/32-inch _____

    d.   3/16-inch _____

48. The head angle of an MS20426 rivet is _____ degrees.

49. A piece of 0.040-inch aluminum alloy skin should be _____ (dimpled or countersunk) for the installation of an MS20426AD5 rivet.

50. When three sheets of metal are dimpled and stacked to be riveted together, they should be _____ (coin or radius) dimpled.

51. What type of dimpling is used on magnesium sheets to prevent their cracking in the dimpling process?

    _____ dimpling

52. An AN470DD6 rivet would be most properly driven with a _____ (fast hitting or one-shot) rivet gun.

53. A bucking bar weighing about _____ to _____ pounds should be used when driving 1/8-inch rivets.

54. Indicate the dimensions of a properly driven rivet:

    A = _____D

    B = _____D

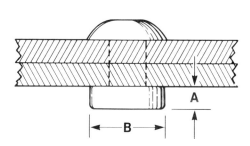

55. When installing rivets according to the N.A.C.A. method, the manufactured head is placed on the _____ (inside or outside) of the structure.

56. The tapping code used in team riveting is:

   a.  Bad rivet                         _____ tap(s)

   b.  Good rivet                        _____ tap(s)

   c.  Drive the rivet some more         _____ tap(s)

57. When practical, bends should be made _____ (across or with) the grain of the metal.

58. Indicate the minimum bend radius for each of these types of metal:

   a.  0.032-inch 5052-O          _____ inch(es)

   b.  0.040-inch 2024-T4         _____ inch(es)

   c.  0.063-inch 7075-T6         _____ inch(es)

59. The point of intersection of the mold lines of two sides of a bend is called the _____ point.

60. The distance from the mold point to the bend tangent line is known as the

   _____.

61. For a bend of 90 degrees, the setback is calculated using the formula:

   Setback = _____ + _____

62. For a bend of more or less than 90°, we must apply a correction factor known as a(n) _____ -factor to find the setback.

63. The formula used to determine setback of an angle other than 90° is:

   Setback = _____

64. Find the setback for each of these bends:

   a.  T = 0.040-inch          B.R.= 0.250-inch          Degrees = 90

       Setback = _____

b.  T = 0.040-inch          B.R. = 0.250-inch          Degrees = 45

Setback = _____

c.  T = 0.040-inch          B.R. = 0.250-inch          Degrees = 135

Setback = _____

d.  T = 0.125 inch          B.R. = 0.50-inch           Degrees = 90

Setback = _____

65. The amount of material actually involved in the bend is known as the _____
_____.

66. Using the bend allowance chart (figure 5-116) on page 230 of the text, find the
bend allowance for each of these bends.

a.  T = 0.040-inch          B.R. = 0.250-inch          Degrees = 90

B.A. = _____

b.  T = 0.032-inch          B.R. = 0.125-inch          Degrees = 90

B.A. = _____

c.  T = 0.032-inch          B.R. = 0.125-inch          Degrees = 135

B.A. = _____

d.  T = 0.051-inch          B.R. = 0.250-inch          Degrees = 45

B.A. = _____

e.  T = 0.051-inch          B.R. = 0.250-inch          Degrees = 135

B.A. = _____

67. When bending a bulb angle into a convex curve, the flange of the metal must be
_____ (shrunk or stretched).

68. If a piece of metal becomes too hard when it is being bumped into a compound
curve for, it _____ (may or may not) be annealed to soften it.

69. What is done to lightening holes in a wing rib to add stiffness?

They may be _____.

70. Small cracks in the edge of low-stress components such as engine cowling can
usually be _____ at their end to prevent the
crack from enlarging.

71. Shallow scratches in aluminum may be repaired by _____.

72. What form must be completed when returning an aircraft to service following major structural repairs?

    FAA Form _____

73. Why is an octagonal patch preferred over a rectangular patch in a high stressed aircraft structure?

    _____

    _____

74. What may be used to replace a Hi-Shear rivet if you do not have the facilities to install another Hi-Shear rivet?

    _____

75. What diameter mechanical-lock Cherry rivet is normally used to replace an MS20470AD4 rivet?

    _____-inch diameter.

76. If the rivet spacing in the seams of a wing panel are not the same, which spacing should be copied when replacing a section of the panel?

    The nearest seam _____ (inboard or outboard).

77. What tool is used to form a skin path to the contour of the aircraft fuselage skin?

    _____

78. When repairing aircraft floats, the rivets should be dipped in sealant and driven while they are _____ (wet or dry).

# Chapter VI
# Aircraft Wood And
# Composite Structural Repair

1. To prevent wood structure rotting, it should be treated with a(n) _____ sealer.

2. The best finish for internal wood structure is a good coat of _____ _____.

3. _____ plywood, covered with aircraft fabric may be used as the external skin on some aircraft.

4. The _____ (lowest or highest) points inside an aircraft structure are the most likely places for wood deterioration to begin.

5. If wood comes up on small chunks rather than splinters when it is picked with the point of a knife blade, it may be considered to be _____ (good or rotted).

6. The reference wood for aircraft structure is _____.

7. Two commonly used types of glue for modern wood aircraft construction are:
   a. _____
   b. _____

8. Sandpaper _____ (should or should not) be used to prepare a surface for gluing.

9.  Almost all adhesives have three time periods that are critical to their use, they are:

    a.  _____

    b.  _____

    c.  _____

10. A wing spar _____ (may or may not) be spliced under the fitting for a lift strut.

11. Dimension this scarf splice in a solid spruce wing spar, using the recommended dimensions.

    a.  Dimension A _____

    b.  Dimension B _____

    c.  Dimension C _____

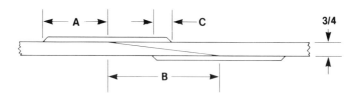

12. The largest hole in a plywood skin that can be repaired using a doped-on cloth patch is _____ inch(es) in diameter.

13. A splayed patch cannot be used in plywood skin that is more than _____- inch(es) thick.

14. The taper for a splayed patch is _____:1.

15. The taper for a scarfed patch in a plywood skin is _____:1.

16. Plastics, generally, may be classified as one of these two types:

    a.  _____

    b.  _____

17. _____ (Thermoplastic or Thermosetting) resins have very little strength in themselves, and are usually reinforced with paper, cloth, or other filaments.

18. These four groups of resins make up the plastic materials with which we are familiar:

    a. _____

    b. _____

    c. _____

    d. _____

19. The pulleys used in aircraft control systems are usually made of _____ cloth impregnated with a(n) _____ resin.

20. The two common types of fiberglass are:

    a. _____

    b. _____

21. _____ is the name given to aromatic polymide fibers, such as Kevlar.

22. Carbon fiber is stronger in _____ (compressive or tensile) strength than Kevlar.

23. Firewalls may be made of _____ fiber composites to dissipate the heat.

24. The strength of a fiber is _____ (parallel or perpendicular) to the direction that the threads run.

25. The threads which run the length of the fabric as it comes off the bolt are referred to as the _____ (warp or weft).

26. The tightly woven edge produced by the weavers to prevent the edges from ravelling is referred to as the _____ edge.

27. The _____ is at a 45° angle to the warp threads.

28. Material in which all of the major fibers run in one direction, giving strength in that direction, are known as _____.

29. Chopped fibers that are compressed together are often called _____.

30. Fabrics are _____ (more or less) resistant to fiber breakout, delamination, and more damage tolerant than unidirectional materials.

31. Five manufacturing methods that may be used with laminated construction are:

    a. _____

    b. _____

    c. _____

    d. _____

    e. _____

32. What five methods are used to dissipate the electrical charge on composite structures?

    a. _____

    b. _____

    c. _____

    d. _____

    e. _____

33. A(n) _____ is added to polyester resin to prevent its hardening before it should.

34. When catalysts and accelerators are mixed together, _____ is generated.

35. Accelerator must always be added to the _____ - never to the catalyst alone.

36. A _____ (thick or thin) layer of catalyzed polyester resin will cure faster.

37. Polyester resin will shrink _____ (more or less) than epoxy resin.

38. Unmixed polyester resin should be stored at a temperature between _____ and _____ degrees Fahrenheit.

39. Technically, epoxy does not use a catalyst, as polyester resin does, it uses a(n) _____ agent which combines with the epoxy.

40. _____ are a popular thixotropic agent for use with thermosetting resins to give the resin good body with a minimum of weight.

41. Pre-impregnated fabrics, or pre-pregs, are simply fabrics that have the _____ system already impregnated into the fabric.

42. In addition to microballoons, _____ and _____ may be used as filler material when repairing composite structures.

43. Honeycomb core material may be constructed of:

    a. _____

    b. _____

    c. _____

    d. _____

    e. _____

    f. _____

    g. _____

44. Closed cell styrofoam should be used with _____ (epoxy or polyester) resin only.

45. _____ foam can be used with either epoxy or polyester resin.

46. Damage to aircraft composite structure can usually be placed in one of these three categories:

    a. _____

    b. _____

    c. _____

47. To detect internal flaws or areas of delamination a(n) _____
    _____ test may be used.


48. _____ locates flaws by temperature variations at the surface of
    a damaged part.


49. Radiography (X-ray) _____ (may or may not) be used to detect water
    inside honeycomb core cells.


50. The older type fiberglass repairs _____ (can or cannot) be used on
    advanced composite structures.


51. Air-driven _____ are the best tools to use to remove damaged
    honeycomb core material from an aircraft structure.


52. _____ is a recommended solvent to clean the surface of a fiber-
    glass structure before laying in a bonded repair.


53. _____ is the separation of layers of material in a laminate.


54. A dent in a honeycomb panel _____ (will or will not) reduce the
    strength of the panel.


55. If a puncture in the metal face of a honeycomb structure is minor (less tan 1-inch),
    the repair may be made by _____.


56. If a damaged area is too large to be repaired by a potted repair, it may be cleaned
    out and filled with a plug made of honeycomb material or _____
    _____.


57. When riveting a repair to a honeycomb structure, the rivets should be dipped in
    _____ and upset while they are wet.

58. When repairing a radome, you must be sure that three characteristics are considered. These are:

    a. _____

    b. _____

    c. _____

59. Repairs made to a structural rib are usually considered to be _____ (temporary or permanent).

60. Repairs made to advanced composites using materials and techniques that have traditionally been used for fiberglass repairs will result in an _____ (airworthy or unairworthy) repair.

61. _____ is the method that is usually used to remove advanced composite plies with the most control.

62. _____ cutting is used to remove damaged material with a tapered cutout.

63. The water break test is used to detect _____ or _____ _____ contamination.

64. The _____ of honeycomb is the direction of the honeycomb that can be pulled apart.

65. A(n) _____ is a tool which can be used to reference the orientation of the warp of the fiber.

66. If lightening protection is installed during a composite repair, it may be tested for continuity using a(n) _____.

67. Mechanical pressure is used during the curing operation to:

    a. _____

    b. _____

    c. _____

    d. _____

    e. _____

68. _____ is probably the most effective method to apply pressure to a repair.

69. A barrier between the wet patches and the other bagging materials may be provided by using _____ fabrics and films.

70. _____ are absorbent materials which are used to soak up the excess resins.

71. Composite matrix systems may be divided according to type of curing, they will be one of these two types:

    a. _____

    b. _____

72. _____ curing is the process of raising the temperature to a point, holding it there, then raising the temperature, holding it at the new value, and continuing this process until the cure temperature is reached. After the cure time has elapsed, the reverse procedure will be used to slowly cool the material.

73. The use of heat lamps to cure composite parts _____ (is or is not) recommended.

74. An autoclave provides both _____ and _____ for curing composite materials.

75. Heating _____ is/are probably the most widely accepted form of applying heat to a composite component for repair work.

76. Scissors with special steel blades with serrated edges are used to cut through _____ fabric.

77. You _____ (should or should not) use a cutting coolant when drilling into bonded honeycomb or foam core structure.

78. Whenever possible composite materials should be _____ with wood when drilling.

79. Twist drills used on composite materials should have an included angle of
_____ (how many) degrees, and be used in a _____ (high or low)
speed drill motor.

80. Special brad point drills are available for drilling _____ (what
type) composites.

81. Aluminum fasteners _____ (should or should not) be used with car-
bon/graphite material.

82. Aluminum oxide sandpaper _____ (should or should not) be used to
sand carbon/graphite materials.

83. Cutting surfaces of tooling used on composites should be _____
(what material) coated whenever possible.

84. Composite materials which have exceeded their storage life should be _____
_____.

85. While sanding, drilling, or trimming composite materials _____
must be worn to prevent breathing toxic fumes or dust.

86. The two most frequently used transparent plastics for aircraft windows and windshields are:

   a. _____

   b. _____

87. Acetone _____ (will or will not) soften acrylic plastic.

88. The most satisfactory way to store acrylic sheets is _____ (a, b, or c)

   a.   Store them horizontally

   b.   Store them vertically

   c.   Store them slightly away from the vertical

89. When a plastic material has thousands of tiny cracks in its surface, it is said to be
_____.

90. Soaking a sheet of acrylic plastic in boiling water is a(n) _____ (satisfactory or unsatisfactory) way of heating it to form it.

91. The included angle for a twist drill used to drill acrylic plastic should be about _____ degrees.

92. _____ is a good material for softening acrylic plastic to glue it.

93. A cemented joint in acrylic plastic _____ (does or does not) shrink as it becomes hard.

94. Curing acrylic plastics at an elevated temperature is also referred to as _____ _____.

95. Acrylic windshields may be cleaned with _____ and water.

# Chapter VII
# Aircraft Welding

1. _____ welding joins metals by blending compatible molten metals into one common part or joint.

2. _____ or nonfusion joining occurs when two or more pieces of steel are held together by a noncompatible molten brass or silver material.

3. Brazing materials melt at temperatures above _____ degrees Fahrenheit, while solders melt at temperatures below this.

4. The three basic types of welding are:

    a. _____

    b. _____

    c. _____

5. Gas welding utilizes what two gases?

    a. _____

    b. _____

6. Shielded Metal Arc Welding (SMAW) is more commonly known as _____ _____ welding.

7. MIG welding may also be known as _____ arc welding.

8. The form of arc welding that fills most of the needs in aircraft maintenance is known as _____ arc welding. This is commonly referred to by what three letter designation? _____

9. An electric arc produces a temperature in excess of _____ degrees Fahrenheit.

10. An uncoated wire is fed into the torch as a consumable electrode in _____ (TIG or MIG) welding.

11. Low-voltage, high-current _____ (DC or AC) power is used for MIG welding.

12. A small wire of _____ is used as the non-consumable electrode for TIG welding.

13. Two forms of electrical resistance welding are:

    a. _____

    b. _____

14. What three variables are controlled to get the proper weld using either of the two electrical resistance welding methods?

    a. _____

    b. _____

    c. _____

15. Acetylene gas stored under a pressure of more than _____ pound(s) per square inch will become unstable.

16. Normal operating pressure for most welding jobs using acetylene is _____ to _____ psi.

17. Acetylene gas may be stored under pressure by dissolving it in _____.

18. The amount of acetylene gas in a steel cylinder is determined by the _____ (weight or pressure) of the cylinder.

19. The acetylene tank valve should be opened no more than _____ to _____ turn.

20. The main advantage to oxyhydrogen welding is that the hydrogen flame burns much _____ (cleaner or hotter) than acetylene.

21. Oxygen must never be used in the presence of _____ base substances.

22. The temperature of a neutral oxyacetylene flame is about _____ degrees Fahrenheit.

23. The union nut for connecting the oxygen regulator to the cylinder valve has _____ (right or left) hand threads.

24. The acetylene regulator has _____ (right or left) hand threads.

25. Turning the regulator adjusting handle clockwise will _____ (open or close) the valve.

26. The acetylene hose is colored _____ and has a _____ (left or right) hand threaded fitting at each end.

27. The oxygen hose is colored _____ and has a _____ (left or right) hand threaded fitting at each end.

28. The fittings on the _____ (oxygen or acetylene) hose are identified by a groove cut around their hexes.

29. The torch most often used with cylinder gases is an _____ (balanced-pressure or injector)-type.

30. A torch designed for light-duty welding, such as that of aircraft thin wall tubing, has its valve on the end of the torch near the _____ (tip or hoses).

31. The size of the orifice in a welding tip is measured with the shank of a(n) _____.

32. The lower the number of the welding tip (not the size of the orifice), the _____ (smaller or larger) the tip.

33. The lower the number of the welding goggle filter, the _____ (lighter or darker) the filter.

34. A _____ (blue or brown) filter should be used in the welding goggles for welding aluminum.

35. Oxyacetylene welding goggles _____ (may or may not) be used while arc welding.

36. Most mild steel filler rods are coated with _____ to prevent rust from forming on the surface of the rod.

37. The _____ of the metal to be welded will determine the size of the torch tip needed.

38. An oxyacetylene flame in which the feather of the outer cone just disappears into the inner cone is a(n) _____ (oxidizing, neutral, or carburizing) flame.

39. A rosette weld is a form of _____ (lap or butt) weld.

40. A cleanly formed oxyacetylene weld should encompass the following qualities:

    a. _____

    b. _____

    c. _____

    d. _____

41. Penetration of welded aircraft parts should be at a depth of _____ percent of the thickness of the base metal.

42. Which gas is better for torch welding, aluminum, acetylene or hydrogen?

    _____

43. Aluminum _____ (does or does not) change its color as it melts.

44. The flame used for welding aluminum should be neutral or slightly _____ (carburizing or oxidizing).

45. The pre-heat flame in an oxyacetlyene cutting torch should be adjusted to a(n) _____ (neutral or oxidizing) flame.

46. In brazing the filler metal is pulled between closely fitting parts by
    _____ action.

47. When brazing, the filler rod should be melted by _____ (a or b).
    a.  holding the rod in the flame of the torch.
    b.  touching the rod to the hot metal.

48. Soft solder is a mixture of _____ and _____.

49. Fittings may be attached to stainless steel oxygen lines by _____
    soldering.

50. An _____ shields the puddle of molten
    metal to prevent the formation of oxides when TIG welding.

51. When using direct current for TIg welding, the most heat is put into the work when
    the DC is connected with _____ (straight or reverse) polarity.

52. The tungsten electrode should be _____ (pointed or rounded) for
    straight-polarity DC welding.

53. Aluminum should be TIG welded using _____ (AC or DC).

54. The end of the tungsten electrode used with AC should be _____
    (pointed or rounded).

55. Exhaust stacks from an aircraft engine should be TIG welded using _____ (DC or AC).

56. Highly stressed steel tubing used in aircraft structure is most likely SAE
    _____ steel.

57. Advisory Circular _____ shows typical repair schemes for
    welded repairs of steel tube structure.

58. The ends of a reinforcing tube used over a damaged tube in an aircraft structure
    can be cut with either a 30-degree _____ or a(n)
    _____.

# Chapter VIII
# Ice And Rain Control Systems

1.  Frost must be removed from the wings of an airplane before flight because it forms an effective aerodynamic _____ on the surface, and increases drag.

2.  There _____ (does or does not) have to be visible water in the air for carburetor ice to form.

3.  The formation of ice is prevented by _____ (anti-icing or deicing) systems.

4.  Ice is removed from aircraft surfaces by _____ (anti-icing or deicing) systems.

5.  The three types of anti-icing systems are:

    a.  _____

    b.  _____

    c.  _____

6.  Turbine powered aircraft may use _____ air to heat the leading edges of the wings to prevent the formation of ice.

7.  The inlet guide vanes on a turbine engine are provided with anti-icing protection from _____ (bleed air or electrical current).

8.  Pitot heads are protected from icing with _____ (hot air or electrical heaters).

9.  If there is no provision for ice removal on the static port, the aircraft should be equipped with an _____ air source valve.

10. Windshields are normally _____ (electrically or hot air) heated to prevent ice from obstructing the flight crew's vision.

11. An aircraft windshield has a greater bird-impact resistance when it is _____ (heated or cold).

12. Chemical anti-icing systems may be found on these components:

    a. _____

    b. _____

    c. _____

13. Propeller anti-icing systems normally use _____ alcohol to prevent icing.

14. The air to inflate rubber deicer boots installed on an airplane powered by a reciprocating engine is taken from the discharge of the _____ pump.

15. When the tubes in deicer boots are deflated, they are held tightly against the wing by _____.

16. Small turbine powered aircraft may utilized pneumatic deicing boots that are inflated using _____ air.

17. Two ways of securing rubber deicer boots to the leading edge of a wing are:

    a. _____

    b. _____

18. The surfaces of rubber deicer boots should be kept clean by periodically washing them with a solution of _____ and _____.

19. Propellers are normally deiced with _____ (hot air or electrothermal boots).

20. Electrical current is supplied to the propeller deicers using _____ and _____ assemblies.

21. Electric propeller deicers operate _____ (continuously or on a sequenced cycle).

22. Frost may be removed from an aircraft on the ground by spraying it with a mixture of _____ and _____ _____.

23. Three methods of rain control are:

    a. _____

    b. _____

    c. _____

24. Two methods of operating aircraft windshield wipers are:

    a. _____

    b. _____

25. Chemical rain repellent should be used only in a _____ (heavy or light) rain.

26. Pneumatic rain removal systems use _____ air from the engine.

# Chapter IX
# Hydraulic And Pneumatic
# Power Systems

1.  Hydraulic system have many advantages, including:

    a. _____

    b. _____

    c. _____

    d. _____

2.  Hydraulic systems are almost _____% efficient.

3.  Hydraulic fluid is considered to be _____ (compressible or incompressible).

4.  The pressure exerted by a column of liquid is determined by the _____ _____ (height of the column or the volume of the container).

5.  Force per unit area is a measure of _____.

6.  The relationship between force, area and pressure may be expressed using the formula _____.

7.  In a hydraulic system the relationship between the area of the piston, the distance it moves, and the volume of the fluid displaced may be expressed using the formula:

    _____

8.  a.   If $F_1$ is 100 pounds, $F_2$ will be _____ pounds.

    b.   If the piston in cylinder-1 moves 5 inches, the piston in cylinder 2 will move _____ inches.

    c.   A force of _____ pounds will have to be exerted on the piston in cylinder 1 to produce a force of 1,000 pounds on the piston in cylinder 2.

    d.   If the pressure inside cylinder 1 is 500 psi, the pressure inside cylinder 2 will be _____ psi.

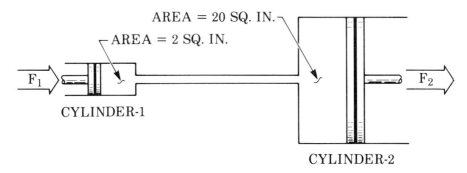

9.  a.   If a pressure of 500 psi is applied to both sides of the piston in this cylinder, the piston will move _____ (up or down).

    b.   For force produced by the piston will be _____ pounds.

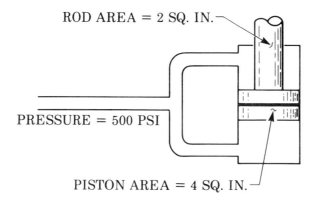

10. Fill in the blanks with the missing values:

|     | PRESSURE | AREA | FORCE | DISTANCE | VOLUME |
|-----|----------|------|-------|----------|--------|
| a.  | 500 psi | 10 sq. in. | _____ pounds | | |
| b.  | 1,000 psi | _____ sq. in. | 5,000 pounds | | |
| c.  | _____ psi | 6 sq. in. | 24,000 pounds | | |
| d.  | | 12 sq. in. | | 6 inches | _____ cu. in. |
| e.  | | 25 sq. in. | | _____ in. | 1,000 cu. in. |
| f.  | | _____ sq. in. | | 15 inches | 500 cu. in. |
| g.  | _____ kg/sq. cm. | 1,000 sq. cm. | 100 kg | | |
| h.  | 500 kg/sq. cm. | _____ sq. cm. | 200 kg. | | |
| i.  | 100 kg/sq. cm. | 25 sq. cm. | _____ kg. | | |
| j.  | | 450 sq. cm. | | _____ cm. | 1,800 cu. cm. |
| k.  | | _____ sq. cm. | | 25 cm. | 2,500 cu. cm. |
| l.  | _____ kg/sq. cm. | 45 sq. cm. | 50 kg | | |
| m.  | 2,000 kg/sq. cm. | _____ sq. cm. | 40 kg | | |
| n.  | 1,000 kg/sq. cm. | 500 sq. cm. | _____ kg | | |
| o.  | | 100 sq.cm. | | _____ cm. | 600 cu. cm. |
| p.  | | 500 sq. cm. | | 1 meter | _____ cu. cm. |

11. A good hydraulic fluid must posses these properties:

a. _____

b. _____

c. _____

d. _____

e. _____


12. _____ is the internal resistance to flow of a liquid.


13. _____ is the temperature at which a liquid will give off vapor in sufficient quantity to ignite momentarily.


14. The three types of hydraulic fluid currently being used in civilian aircraft are:

a. _____

b. _____

c. _____


15. Vegetable base hydraulic fluid is essentially _____ oil and

_____.


16. Vegetable base hydraulic fluid is dyed _____ for identification.


17. The most widely used hydraulic fluid in general aviation aircraft today is

_____.


18. What type of hydraulic fluid is dyed red for identification?

_____


19. Synthetic hydraulic fluids were developed to provide a(n) _____-resistant hydraulic fluid for use in high performance piston and turbine aircraft.


20. Skydrol is _____ in color.

21. Name the fluid that may be used to flush hydraulic systems using each of the following types of hydraulic fluid:

    a.   Vegetable-base    _____

    b.   Mineral-base      _____

    c.   Synthetic         _____


22. Name the material used for seals in hydraulic systems using each of the following types of hydraulic fluids:

    a.   Vegetable-base    _____

    b.   Mineral-base      _____

    c.   Synthetic         _____


23. If Skydrol is accidently spilled, it should be immediately cleaned up using _____ _____ and _____.


24. A hydraulic system must contain three basic types of units; these are:

    a.   _____

    b.   _____

    c.   _____


25. A single unit containing an electrically driven hydraulic pump, reservoir, control valve and many of the auxiliary valves is known as a hydraulic _____ _____ system.


26. The two types of hydraulic reservoirs are:

    a.   _____

    b.   _____


27. Aircraft that operate in the lower altitudes normally use _____ (pressurized or unpressurized) hydraulic reservoirs.


28. Three ways of pressurizing a hydraulic reservoir are:

    a.   _____

    b.   _____

    c.   _____

29. One micron is equal to _____ inch(es).

30. Pleated paper micronic filters are usually installed in the hydraulic system _____ (pressure or return) line.

31. Cuno filters _____ (can or cannot) be used on the pressure side of the hydraulic system.

32. A _____ (constant or variable) displacement pump moves a specific volume of fluid each time its shaft turns.

33. A gear-type hydraulic pump is a _____ (constant or variable) displacement pump.

34. An _____ valve of some sort is needed when a constant displacement pump is used.

35. A hydraulic systems using a variable displacement pump _____ (does or does not) require an unloading valve.

36. The _____ (open or closed) -center valve directs fluid through the center of the valve back to the reservoir when a unit is not being actuated.

37. A valve that allows free flow of fluid in one direction, but no flow in the opposite direction is called a(n) _____ valve.

38. A valve that allows free flow of fluid in one direction but a restricted flow in the opposite direction is called a(n) _____ valve.

39. A valve that requires one component to fully actuate before another can actuate is called a(n) _____ valve.

40. _____ valve are similar to sequence valves except they are opened by hydraulic pressure rather than by mechanical contact.

41. A(n) _____ may be installed to block a line in case of a serious leak.

42. Two principles upon which a hydraulic fuse may operate are:

    a. _____

    b. _____

43. The simplest type of pressure control valve is the _____ valve.

44. A(n) _____ may be installed when it is
    necessary to operate some portion of a hydraulic system at a pressure lower than
    the normal system pressure.

45. Three type of accumulators used in aircraft hydraulic systems are:

    a. _____

    b. _____

    c. _____

46. One compartment of an accumulator is conneced to the hydraulic system, and the
    other compartment is filled with compressed _____ or
    _____.

47. High pressure valve cores are identified by the letter _____ embossed on the
    end of their stem.

48. Hydraulic actuator which produce straight-line movement are known as
    _____ actuators.

49. If a continuous rotational force is needed, a hydraulic _____
    may be used.

50. A chevron seal is a _____ (one-way or two-way) seal.

51. An O-ring is a _____ (one-way or two-way) seal.

52. An O-ring should have a width of approximately 10% _____ (larger or
    smaller) than the depth of the groove into which it fits.

53. A backup ring should be installed on the side of an O-ring _____ (toward or away from) the pressure.

54. The _____ date is the date of O-ring manufacture.

55. An O-ring marked with a blue dot _____ (would or would not) be compatible with MIL-H-5606 hydraulic fluid.

56. The air for _____ (high, medium, or low) pressure pneumatic systems is usually stored in metal bottles.

57. Medium pressure pneumatic systems on turbine powered aircraft usually use _____ air for their operation.

58. _____ (Vane or Piston)-type pumps are usually associated with low pressure pneumatic systems, such as those used to operate aircraft instruments.

59. After the air in a pneumatic system leaves the moisture separator it must pass through a(n) _____ or _____ _____ to remove the last traces of moisture.

60. A(n) _____ valve may be installed to allow a pneumatic system to operate from a ground source.

61. Emergency extension air enters a hydraulic landing gear actuator through a(n) _____ valve.

# Chapter X
# Aircraft Landing
# Gear Systems

1. Retracting the landing gear on an aircraft decreases the _____ (induced or parasite) drag.

2. The tailwheel type landing gear arrangement is also known as _____ _____ landing gear.

3. Almost all current production airplanes use the _____-type landing gear arrangement.

4. The streamlined fairings used to enclose the wheels are referred to as _____ _____.

5. A(n) _____ cord is a bundle of small strands of rubber encased in a loosely woven cloth tube.

6. The most widely used shock absorber for aircraft is the _____ _____ shock absorber or more commonly known as an _____ strut.

7. An oleo strut absorbs the landing impact shock with the _____ (air or oil).

8. An oleo strut absorbs the taxi shocks with the _____ (air or oil).

9. Oleo strut inflation is checked by measuring the _____.

10. If the front sides of the wheels of an aircraft are closer together than the rear sides, the landing gear is _____ (toed-in or toed-out).

11. If the tops of the wheels of an aircraft are closer together than the bottoms, the landing gear has a _____ (positive or negative) camber.

12. A(n) _____ switch may be installed in a landing gear retraction system to prevent the landing gear from being raised while the aircraft is on the ground.

13. A gear warning horn will sound if the throttles are _____ (advanced or retarded) and the landing gear is in the _____ (up or down) position.

14. A(n) _____ is used to reduce the back and forth oscillations of the nose wheel at certain speeds.

15. Aircraft wheels may be cast or forged of either _____ or _____ _____ alloy.

16. The _____ area is the most critical part of a wheel.

17. One or more _____ may be installed in the inboard half of the main wheels of jet aircraft to release the air from the tire in case of extreme overheating.

18. Aircraft bearings _____ (should or should not) be cleaned with steam.

19. Water stains on a wheel bearing is cause for rejection, because this may be an indication of _____ corrosion.

20. Discolored bearings indicate damage due to _____.

21. Dye penetrant inspection _____ (is or is not) a good method to use to locate cracks in the bead seat area of an aircraft wheel.

22. Wheel bolts should be inspected using the _____ _____ inspection method.

23. If _____ (how many) of the fusible plugs in the wheel show any signs of softening, they must all be replaced.

24. Power brakes are operated using pressure from the _____ hydraulic system.

25. The hydraulic pressure to the brake system is reduced using a(n) _____ _____ system.

26. In case of total failure of the hydraulic system, most large aircraft brake systems can be operated using a(n) _____ emergency system.

27. The amount the automatic adjuster pin protrudes from the cylinder head of a Goodyear single-disk brake may be used as an indicator or the wear of the _____ (linings or disk).

28. Cleveland brake linings should be replaced when they have worn to a thickness of _____ inch(es).

29. _____ brake action is nearly always an indication of air in the system.

30. The operation performed to remove air from the brake system is known as _____ _____ the brakes.

31. When bleeding air from a master cylinder brake system using the pressure method, the fresh fluid is introduced at the _____ (wheel cylinder or reservoir).

32. If a brake system is inadvertently serviced with the wrong type of hydraulic fluid what must be accomplished?

    a. _____

    b. _____

33. During brake overhaul the housings should be checked using the _____ _____ inspection method.

34. It _____ (is or is not) necessary to remove the wheel to replace the linings on a Cleveland single-disk brake.

35. Anytime a brake shows signs of _____ it must be removed from the aircraft, inspected, and rebuilt.

36. When a brake fails to completely release after the pressure is removed, it is said to be _____.

37. The three basic components of a brake anti-skid system are:

    a. _____

    b. _____

    c. _____

38. The three main functions of an anti-skid brake control box are:

    a. _____

    b. _____

    c. _____

39. The test lights in an anti-skid system are ON when there _____ (is or is not) pressure in the system.

40. The most popular low-pressure tire found on piston-powered aircraft today is the Type _____.

41. The ply rating _____ (does or does not) indicate the actual number of fabric plies in the tire.

42. Tubeless tires are identified by the word _____ on their sidewall.

43. Aircraft tires are designed to deflect _____ (more or less) than automobile tires.

44. The most popular tread pattern found on aircraft tires today is the _____ _____ tread.

45. A tire with a chine, or deflector, on its sidewall is used on the _____ (nose or main) wheel.

46. The most important aspect of tire preventive maintenance is maintaining _____ _____.

47. Identify the inflation condition that has caused the wear pattern on each of these aircraft tire sections:

    1. Under inflation      2. Over inflation      3. Correct inflation

a.

b.

c.

48. The greatest heat will be generated in a tire that is operated _____ (under-inflated or over-inflated).

49. The inflation pressure specified by the _____ (tire or airframe) manufacturer should be used to determine the proper inflation of a tire.

50. Air pressure should be checked when the tire is _____ (hot or cold).

51. Retreading _____ (is or is not) a recommended procedure for aircraft tire operation.

52. What type of seal is used between the halves of a wheel using a tubeless tire? A(n) _____ seal.

53. Cracks in the sidewall rubber that expose the body cords of the tire _____ (are or are not) cause for rejection of the tire.

54. If any of the bead area is damaged, the tire _____ (can or cannot) be repaired.

55. There _____ (is or is not) a limit on the number of times an aircraft tire can be retreaded.

56. Ozone _____ (is or is not) harmful to rubber products.

57. Tires should be stored _____ (vertically or horizontally) whenever possible.

58. The two primary causes for an aircraft tube leaking are:

    a. _____

    b. _____

59. If a tube is suspected of leaking, first check the _____.

60. The red dot on an aircraft tire generally indicates the _____ (light or heavy) spot.

61. Wheel half bolts usually require an _____ compound be applied to the threads.

62. The yellow mark on an aircraft tube usually indicates the _____ (light or heavy) point.

63. Three ways of attaching balance weights to an aircraft wheel are:

    a. _____

    b. _____

    c. _____

64. _____ is the controlled movement of the airplane under its own power, while on the ground.

65. _____ occurs when the aircraft's tires lose contact with the taxiway or runway surface, generally because of standing water on the surface.

# Chapter XI
# Fire Protection Systems

1. For a fire to occur, there must be:

    a. _____

    b. _____

    c. _____

2. Name the NFPA class of fire that is described by each of these conditions:

    a. A fire in which there are energized electrical circuits is a class-_____ fire.

    b. A fire with liquid fuel is a class-_____ fire.

    c. A fire in which there is burning metal is a class-_____ fire.

    d. A fire with solid combustibles as the fuel is a class-_____ fire.

3. The thermal switch fire detection system is a(n) _____-type system.

4. Thermal switch fire detection units respond to a _____ (pre-set temperature or a rate-of-temperaure-rise).

5. Thermal switch fire detectors are connected in _____ (series or parallel) with each other.

6. Thermocouple fire detection systems may also be known as the _____ _____ fire detection system.

7. Thermocouple fire detection systems respond to a _____ (pre-set temperature or a rate-of-temperature-rise).

8. The two relays in a thermocouple fire detection system are:

    a. _____

    b. _____

9. The thermocouples in an Edison fire detection system are connected in _____ (series or parallel) with each other.

10. In the thermocouple system, one thermocouple called the _____ thermocouple is placed in a location that is relatively well protected from the initial flame.

11. More complete coverage of a fire hazard area than is provided by a spot-detector system may be obtained using a(n) _____ _____ fire detection system.

12. The _____ (Fenwall or Kidde) continuous loop system uses two wires inside and inconel tube.

13. A _____ (sensor-responder or continuous loop) fire detection system may also function as an overheat indicator.

14. The two types of sensor-responder systems that may be found in use are:

    a. _____

    b. _____

15. The wing compartments and wheel wells would be considered a class _____ fire zone.

16. Three types of smoke detection units which may be found on aircraft are:

    a. _____

    b. _____

    c. _____

17. _____-type smoke detectors use a small amount of radioactive material to ionize some of the oxygen and nitrogen drawn into the unit.

18. Halogenated hydrocarbons are identified through a system of _____ _____ numbers.

19. A specially formulated _____ extinguishing agent is used for class D fires.

20. The four types of hand-held fire extinguishers that are usable for aircraft cabin fires are:

    a. _____

    b. _____

    c. _____

    d. _____

21. _____ is used to propel water from a pressurized water type fire extinguisher.

22. The two basic categories into which installed fire extinguisher systems are placed are:

    a. _____

    b. _____

23. If an installed fire extinguishing system has been discharged by normal operation the _____ (red or yellow) blowout disk will be ruptured.

24. If a discharge cartridge is removed from a discharge valve, it _____ (may or may not) be used in another discharge valve assembly.

25. Almost all type of fire extinguisher containers require _____ to determine the state of charge.

26. Fire extinguisher containers must be hydrostatically tested each _____ years.

# Chapter XII
# Aircraft Electrical Systems

1. Electrical current is measured in _____.

2. The opposition to the flow of electrons is _____ and is measure in _____.

3. Electrical pressure is measured in _____.

4. The basic formula expressing the relationship between electrical flow, resistance and pressure is:

   _____

5. Electrical power is expressed in _____.

6. One horsepower = _____ watts

7. Fill in the blanks in this table:

| VOLTS | AMPS | OHMS | WATTS |
|-------|------|------|-------|
| 6 | 3 | | |
| | 10 | 1.5 | |
| 280 | | | 5,600 |
| | | 4 | 144 |
| 30 | | 7.5 | |
| | 3 | | 360 |

8.  Electricity may be produced by:

    a.  _____

    b.  _____

    c.  _____

    d.  _____

    e.  _____


9.  The magnetic field surrounding these wires will _____ (A or B).

    a.  tend to pull them together

    b.  tend to push them apart

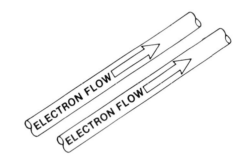


10. What does each of the three digits represent in the left-hand rule for generators?

    a.  First finger      _____

    b.  Second finger     _____

    c.  Thumb             _____

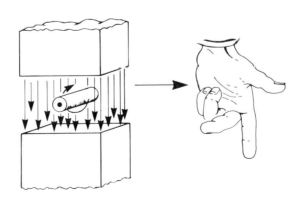

11. What does each of the three digits represent in the right-hand rule for motors?

    a. First finger        _____

    b. Second finger       _____

    c. Thumb               _____

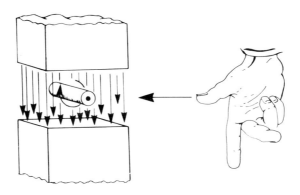

12. Aircraft alternating current electricity has a frequency of _____ hertz.

13. Three oppositions to the flow of alternating current are:

    a. _____

    b. _____

    c. _____

14. The vector sum of the three oppositions to current flow in an AC circuit is known as
    _____ and is represented in a formula by the letter _____.

15. In a purely resistive AC circuit, the voltage and current _____ (are or
    are not) in phase.

16. In a capacitive AC circuit, the changes in current _____ (lead or lag)
    the changes in voltage.

17. In an inductive AC circuit, the changes in current _____ (lead or lag)
    the changes in voltage.

18. The amount of inductive reactance produced by a coil of wire is determined by:

    a. _____

    b. _____

19. The amount of capacitive reactance produced by a capacitor is determined by:

    a. _____

    b. _____

20. Inductive reactance _____ (increases or decreases) as the frequency of the AC increases.

21. Capacitive reactance _____ (increases or decreases) as the frequency of the AC increases.

22. The percentage of current and voltage that are in phase in an AC circuit is called the _____ of the circuit.

23. Write the formula for power in a DC circuit.

    _____

24. Write the formula for true power in an AC circuit.

    _____

25. True power in an AC circuit is expressed in _____.

26. Apparent power in an AC circuit is expressed in _____.

27. The open circuit voltage of a lead acid cell is _____ volts.

28. The specific gravity of a fully charged lead-acid battery is between _____ and _____.

29. The specific gravity of a discharged lead-acid battery is about _____ _____.

30. Nickel-cadmium batteries have a _____ (low or high) internal resistance.

31. The specific gravity of the electrolyte in a nickel-cadmium battery _____ (is or is not) a measure of the state of charge of the battery.

32. To minimize the possibility of "thermal runaway", aircraft that are equipped with nickel-cadmium batteries will have a battery _____ warning system installed.

33. The devices in a DC generator that change the AC in the armature into DC at the output are the _____ and the _____.

34. DC generators employ a _____ relay between the generator output terminal and the bus to prevent the battery from discharging through the generator.

35. Generator output is controlled by varying the _____ current.

36. The three units contained in a three-unit voltage regulator are:

    a. _____

    b. _____

    c. _____

37. Three-unit voltage regulators employ three sets of contact points. Indicate the position of each set of points when the engine is not running (open or closed).

    a. Voltage regulator            _____

    b. Current limiter              _____

    c. Reverse-current cutout relay  _____

38. The alternating current produced in a DC alternator is changed into direct current with a _____ (solid-state or mechanical) rectifier.

39. A reverse-current cutout relay _____ (is or is not) necessary with a DC alternator.

40. Alternators used on large aircraft generally supply _____ (how many) hertz, _____ (how many) phase alternating current.

41. Modern aircraft typically use a _____ (brush-type or brushless) alternator.

42. A _____ unit is used between the engine and the alternator to assure a constant frequency output.

43. A clipping diode may be installed across the coil of the master relay to eliminate _____ that could damage sensitive electronic components.

44. The clipping diode across the coil of the contactor in _____(A or B) is properly installed.

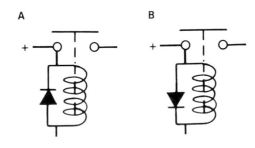

45. The aircraft generator must have sufficient capacity to perform these two functions:

    a. _____

    b. _____

46. The reverse-polarity diode blocks current from flowing to the external power _____ _____ if the applied power is connected backwards.

47. The reverse-polarity diode in _____ (A or B) is properly installed to prevent the external power relay closing if the polarity of the source is incorrect.

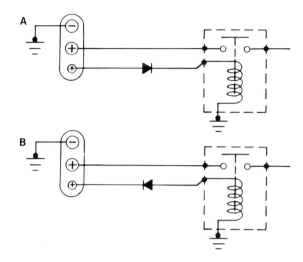

48. The term _____ is often used to define a heavy current solenoid, such as the one used to energize the starter.

49. Identify these statements as True or False concerning the landing gear circuit on page 89. (Note: None of the switches are complete in this drawing.)

    _____    a. All three landing gear must be down and locked before any of the down-and-locked lights will illuminate.

    _____    b. The landing gear selector switch cannot be placed in the GEAR UP position when the weight of the aircraft is on the landing gear.

    _____    c. If the landing gear is not down and locked, the warning horn will sound any time the throttle is closed, regardless of the position of the landing gear switch.

    _____    d. The pump motor cannot run in the GEAR UP direction when the weight of the aircraft is on the landing gear.

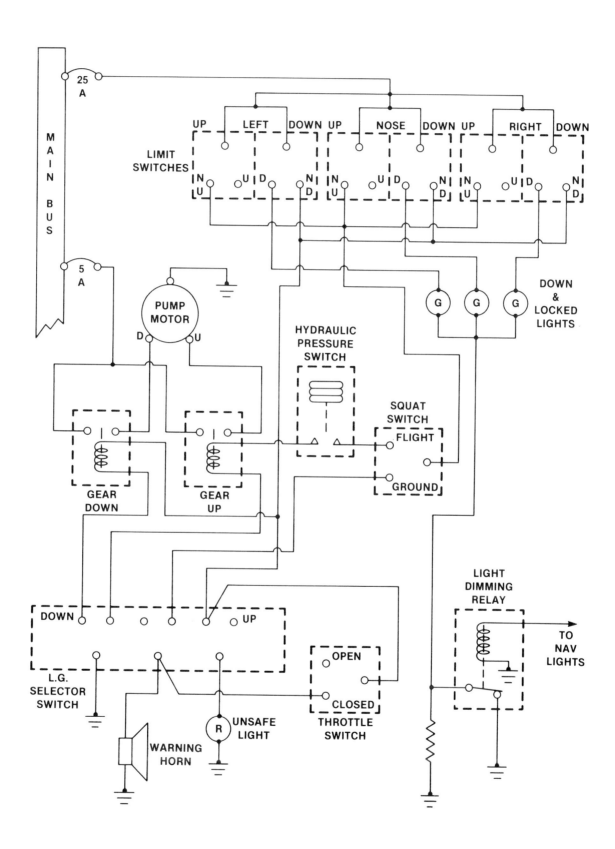

50. Small aircraft, utilizing instrument circuits requiring AC, normally use _____ volt, 400-hertz alternating current.

51. Vibrator type voltage regulators with provisions for paralleling two generators have an extra _____ that is connected to a paralleling switch or relay.

52. The only current that flows in the paralleling coils is that caused by the difference in the output _____ (voltage or current) of the two generators.

53. The two basic types of large aircraft power distribution systems are the:

    a. _____

    b. _____

54. On aircraft equipped with a _____ (split-bus or parallel) system, all generators are connected to a common bus and share the load equally.

55. Identify each of these symbols used in electrical schematic diagrams:

    a.           _____

    b.           _____

    c.           _____

    d.           _____

    e. _____

f. _____

g. _____

h. _____

i. _____

j. _____

k. _____

l. _____

m. _____

n. _____

o. _____

56. The insulation on aircraft wire is designed to withstand _____ volts.

57. _____-gauge wire is the smallest size aluminum wire recommended for use in aircraft electrical circuits.

58. The _____ gauge is used to indicate the size of electrical wire.

59. Encasing a wire in a braided metal jacket is called _____ the wire.

60. A _____ cable consists of a central conductor surrounded by an insulator and a second conductor.

61. Fill in the allowable voltage drop for each of these system conditions:
    a.  Intermittent load in a 28-volt system    _____ volt(s)
    b.  Continuous load in a 14-volt system       _____ volt(s)
    c.  Intermittent load in a 115-volt system    _____ volt(s)
    d.  Continuous load in a 200-volt system      _____ volt(s)

62. Using the electrical wire chart on page 93, determine what gauge wire would be best suited for each of these conditions:
    a.  A continuous load in a 14-volt system needing 40 feet of wire to carry 30 amps. The wire is to be routed in a bundle.

        _____-gauge
    b.  An intermittent load of 200 amps in a 28-volt system. The wire is 12 feet long and routed in free air.

        _____-gauge
    c.  A continuous load of 20 amps in a 115-volt system. The wire is to be routed in a bundle and is 120 feet long.

        _____-gauge
    d.  An intermittent load of 300 amps in a 14-volt system. The wire is 6 feet long and is routed in free air.

        _____-gauge

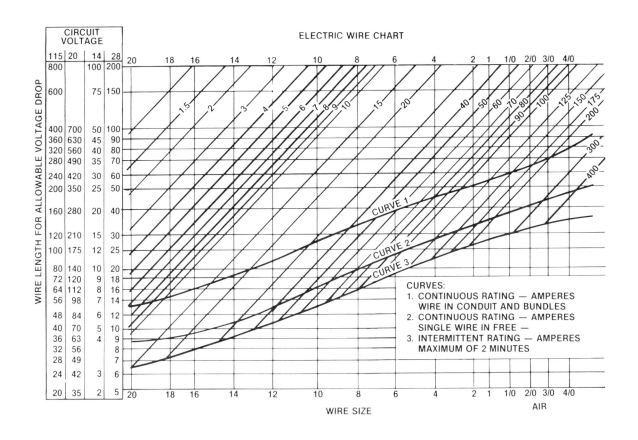

63. When electrical wires are routed parallel to lines carrying oxygen or any type ofliquid, the wiring should be _____ (above or below) the fluid lines.

64. When electrical wiring is run through conduit, the conduit should be approximately _____% larger than the diameter of the wire bundle.

65. Bonding jumpers should have a resistance of no more than _____ ohms.

66. What color insulated terminal would be used on each of these wire sizes:

    a.  20-gauge  _____

    b.  10-gauge  _____

    c.  14-gauge  _____

67. No more than _____ (how many) wire terminals may be connected to a single stud on a terminal strip.

68. The _____ connector is probably the most widely used terminal on coaxial cable attached to aircraft radio antennas.

69. In what Federal Aviation Administration publication may a switch derating factor table be found?

_____

70. A SPST switch has _____ (how many) contacts and controls only one circuit.

71. Normally a _____ (relay or solenoid) has a fixed iron core.

72. A _____ (what type) fuse has a fusible element that is held under tension by a small coil spring.

73. The three basic configurations of circuit breakers used on aircraft are:

   a. _____

   b. _____

   c. _____

74. All aircraft circuit breakers must be of the _____ type.

75. The stationary magnetic field in a DC electric motor may be produced by:

   a. _____

   b. _____

76. Three types of field circuit arrangements used in DC motors are:

   a. _____

   b. _____

   c. _____

77. Series-connected DC motors have a _____ (high or low) starting torque.

78. An aircraft starter uses a _____ (series or shunt)-wound DC motor.

79. The speed of an AC induction motor is determined by:

    a. _____

    b. _____

80. Indicate the color of lens used for each navigation light position listed:

    a.  Left wing          _____

    b.  Right wing         _____

    c.  Tail               _____

81. Anti-collision lights may be either of these two types:

    a. _____

    b. _____

82. Always allow a strobe system to sit in the OFF position for at least _____ minutes prior to any maintenance operation.

# Chapter XIII
# Aircraft Instrument Systems

1.  An aneroid barometer measures _____ (absolute or gauge) pressure.

2.  The two pressures sensed by the airspeed indicator are:

    a. _____

    b. _____

3.  The manifold pressure gauge measures the _____ (absolute or gauge) pressure inside the induction system of a reciprocating engine.

4.  The instrument that measures the differential pressure between the tailpipe total pressure and the compressor inlet total pressure of a turbine engine is the _____ indicator.

5.  The _____ measures the absolute pressure of the air surrounding the aircraft.

6.  The altitude shown on an altimeter when the local altimeter setting is placed in the barometric window is called _____ altitude.

7.  The altitude shown on an altimeter when standard sea level pressure is put into the barometric window is called _____ altitude.

8.  Standard sea level barometric pressure is _____ inch(es) of mercury.

9.  All elevations on aeronautical charts are measured from _____
    _____.

10. Flight level 320 is a pressure altitude of _____ feet.

11. Density altitude _____ (is or is not) a direct measurement.

12. Pressure altitude may be converted into density altitude by correcting the pressure altitude for non-standard _____.

13. Altimeters installed in aircraft that operate under instrument flight rules should have their accuracy checked every _____ (how many) calendar months.

14. The uncorrected reading of an airspeed indicator is called _____ airspeed.

15. The alternate source valve in the aircraft flight instrument system is in the _____ (pitot or static) system.

16. Three instruments that are connected to the static system are:

    a. _____

    b. _____

    c. _____

17. The tests required for a static system of an aircraft operated under instrument flight rules is described in FAR Part _____.

18. On an unpressurized aircraft the static system should not leak more than _____ feet of altitude indication in one minute.

19. The instrument that compares the speed of an aircraft to the speed of sound is known as a(n) _____ meter.

20. Mach _____ is flight at an airspeed of 95% of the speed of sound.

21. The rate-of-climb indicator is more properly known as the _____ _____ indicator.

22. Resistance change-type temperature measuring instruments may use either a(n) _____ or a(n) _____ circuit.

23. A(n) _____ instrument is typically used to measure high temperatures.

24. It _____ (is or is not) permissible to cut off excess length of thermocouple leads.

25. An _____ gives the pilot an indication of the load imposed on the aircraft structure in terms of G's.

26. Alternating current position indictors will be one of these two designs:

    a. _____

    b. _____

27. The _____ (autosyn or magnesyn) uses a permanent magnet for its rotor.

28. The _____ is used by the pilot of a twin-engine aircraft to synchronize the propeller rotations.

29. The cable of a mechanical tachometer operates at _____ (what part) engine speed.

30. _____ tachometers "count" the number of electrical impulses from a special set of contact points located on the magnetos.

31. The two basic characteristics of a gyroscope that makes it useful as a flight instrument are:

    a. _____

    b. _____

32. The artificial horizon indicator senses rotation of an aircraft about which two of its axes?

    a. _____

    b. _____

33. An attitude gyro uses the gyroscopic characteristic of _____.

34. A rate gyro uses the gyroscopic characteristic of _____.

35. The turn and slip indicator is actually two instruments in one housing, the non-gyroscopic instrument is called a(n) _____.

36. The turn and slip indicator senses rotation about the _____ (which axis) axis of the aircraft.

37. The mechanism of a turn coordinator is similar to that used in a turn and slip indicator, except that _____ _____.

38. Meridians of longitude run _____ (north-south or east-west).

39. Because navigation charts are laid out according to the geographic poles and a magnetic compass points to the magnetic poles, we have an error called _____.

40. Lines of equal variation on aeronautical charts are called _____ lines.

41. The line of zero variation is known as the _____ line.

42. Compass _____ (deviation or variation) is caused by the magnetic fields in the aircraft that interfere with those of the earth.

43. Swinging a compass compensates for _____ (deviation or variation).

44. When swinging a magnetic compass the engine _____ (should or should not) be operating.

45. Acceleration and turning error occur only _____ (in-flight or on the ground).

46. The rotor vanes of a dry-type vacuum pump are made of _____ (carbon or steel).

47. Direct current electrical fuel quantity indicators use a variable _____ _____ as a tank unit, or sender.

48. In a capacitance type fuel quantity indicating system the dielectric used is _____.

49. The dielectric value of air is _____.

50. The dielectric value of fuel is approximately _____, but varies according to its temperature.

51. Fuel pressure is typically measure in _____.

52. The mass type flowmeter is the _____ (most or least) accurate method of measuring fuel flow.

53. A(n) _____ is the flight condition where the lift produced by the wings is no longer great enough to hold up the aircraft.

54. The four basic categories of functions of an autopilot are:
    a. _____
    b. _____
    c. _____
    d. _____

55. The two ways an error signal can be generated in an autopilot system are:
    a. _____
    b. _____

56. Dutch roll may be counteracted using a(n) _____ _____.

57. In which position in the basic T configuration for flight instruments are each of these instruments placed?

    a.  Altimeter                    _____

    b.  Attitude gyro                _____

    c.  Directional gyro             _____

    d.  Airspeed indicator           _____

    ② ① ③

    ④

58. Match each of the instrument markings with the condition it represents on an airspeed indicator.

    a.  Never-exceed speed           _____        1.  White arc

    b.  Flap operating speed         _____        2.  Green arc

    c.  Caution area                 _____        3.  Yellow arc

    d.  Best one-engine-out                             4.  Red radial line

        rate of climb speed          _____        5.  Blue radial line

    e.  Normal operating range       _____

59. What is meant by each of these sets of letters?

    a.  EFIS    _____

    b.  CRT     _____

    c.  EADI    _____

    d.  ARINC   _____

    e.  EHSI    _____

60. Greenwich mean time is also known as _____ time.

# Chapter XIV
# Communications And
# Navigation Systems

1.  A potentiometer in a circuit used to dim lights is a form of _____ (analog or digital) electronics.

2.  A switch in a circuit used to turn lights off or on is a form of _____ (analog or digital) electronics.

3.  A diode acts as an electronic _____ valve.

4.  This diode is _____ (forward or reverse)-biased

5.  The three basic components of a common transistor are:
    a. _____
    b. _____
    c. _____

6.  Identify each of the connections to this transistor.

    a.  A is the  _____

    b.  B is the  _____

    c.  C is the  _____

7.  This is a(n) _____(PNP or NPN) transistor.

8.  For a transistor to conduct its emitter-base junction should be _____
    (forward or reverse) -biased.

9.  A pulse of _____ (positive or negative) polarity to the gate is
    needed to start a silicon controlled rectifier conducting.

10. A field-effect transistor is a _____ (low- or high-) impedance device.

11. Identify each of these symbols for electronic devices:

    a.                          _____

    b.  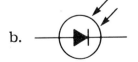                        _____

12. The three basic analog electronic circuits are:

    a. _____

    b. _____

    c. _____

13. A(n) _____ (what type of circuit) converts AC to DC.

14. A(n) _____ (what type of circuit) converts DC to AC.

15. A(n) _____ (what type of circuit) changes the amplitude of the input signal.

16. An electrical or electronic system can perform one or both of these functions:

    a. _____

    b. _____

17. The fundamental building block of digital electronics are _____
    _____.

18. Complete the truth tables for each of these gates:

a. AND Gate

| A | B | C |
|---|---|---|
| — | — | — |
| — | — | — |
| — | — | — |
| — | — | — |

b. OR Gate

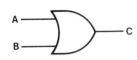

| A | B | C |
|---|---|---|
| — | — | — |
| — | — | — |
| — | — | — |
| — | — | — |

c. NAND Gate

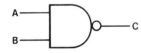

| A | B | C |
|---|---|---|
| — | — | — |
| — | — | — |
| — | — | — |
| — | — | — |

d. NOR Gate

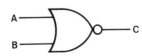

| A | B | C |
|---|---|---|
| — | — | — |
| — | — | — |
| — | — | — |
| — | — | — |

19. Two types of fields exist in a radio transmitting antenna to cause it to radiate electrical energy. They are:

a. _____

b. _____

20. There are two ways a radio carrier wave may be modulated; these are:

    a. _____

    b. _____

21. What is the meaning of the abbreviation IF when discussing super-heterodyne radio receivers?

    _____

22. Most civilian aviation communications is done in the _____ (UHF or VHF) frequency range.

23. VHF and UHF communications and navigation equipment _____ (does or does not) have an operating distance equivalent to the line of sight distance.

24. The length of a transmitting antenna is determined by the frequency for which the antenna is used. The higher the frequency, the _____ (shorter or longer) the antenna.

25. A half-wave antenna may also be referred to as a(n) _____ antenna.

26. Aircraft communications antennas are usually of the _____ (quarter or half) wave type.

27. The _____ (what type) antenna is useful for direction finding.

28. VOR operates in the _____ (VHF or UHF) frequency range.

29. VOR is a _____ (phase or voltage) comparison system.

30. The two pieces of information provided by the VOR indicator, regarding the aircraft's position relative to the course selected are:

    a. _____

    b. _____

31. ADF operates in the _____ (LF/MF or VHF/UHF) frequency range.

32. The two antennas required for the operation of an ADF are:

    a. _____

    b. _____

33. The airborne components of the ILS are:

    a. _____

    b. _____

    c. _____

34. The VOR and the localizer use the same indicator, but the _____ (VOR or localizer) is more sensitive.

35. The Morse code identifier ISAT would identify a _____ (VOR or localizer).

36. A localizer signal is a _____ (phase or voltage) comparison system.

37. A marker beacon uses a single-frequency transmitter with a(n) _____ carrier frequency.

38. The amber light on the marker beacon receiver will light up when the aircraft passes over the _____ (outer or middle) marker.

39. The glide-slope transmitter operates in the _____ (VHF or UHF) frequency range.

40. Tuning to the desired localizer frequency automatically tunes the _____ _____ (marker beacon or glide slope) receiver to the proper frequency.

41. If the aircraft on approach goes above the prescribed altitude, the glide slope needle will point _____ (upward or downward).

42. Distance measuring equipment _____ (is or is not) a form of pulse equipment.

43. The radar beacon transponder operates in the _____ (VHF or UHF) frequency range.

44. It is possible for a pilot to respond with any of _____ (how many) codes on a radar beacon transponder.

45. Altitude information is transmitted from a radar beacon transponder when it is operating in mode _____ (A or C).

46. Two types of radio navigation equipment which may be used to drive the pointers on a Radio Magnetic Indicator (RMI) are:

    a. _____

    b. _____

47. Fill in the color that is used on a color weather radar to indicate the intensity of the rainfall:

    a. Least severe        _____

    b. Medium severity     _____

    c. Most severe         _____

48. Basic weather radar signals are reflected only by the _____ (liquid water or water vapor) inside the storm.

49. _____ (X-band or C-band) radar is better for weather avoidance maneuvers.

50. _____ (X-band or C-band) radar is more common in general aviation airplanes.

51. An RNAV must be tuned to a _____ (VOR or VORTAC) station to operate properly.

52. An emergency locator transmitter operates on these two frequencies:

    a. _____

    b. _____

53. The date of required replacement of ELT batteries must be marked in these two places:

 a. _____

 b. _____

54. Loran receivers operate in the _____ (LF or HF) frequency range.

55. Inertial navigation systems _____ (do or do not) require an external radio signal.

56. _____ (Static wicks or Null field dischargers) use discharge pins to lower the level of static discharge current.

57. Bonding jumpers should have an electrical resistance of no more than _____ ohms.

58. _____ (what type) antenna cable is constructed of a center conductor surrounded by a dielectric which is surrounded by a shield or outer conductor and that is surrounded by an outer protective insulator.

# Chapter XV
# Aircraft Fuel Systems

1.  It _____ (is or is not) permissible for a fuel pump to draw fuel from more than one tank at a time.

2.  A gravity-feed fuel system must be able to flow at least _____ % of the takeoff fuel flow.

3.  A pump-feed fuel system must be able to flow at least _____ % of the takeoff fuel flow.

4.  Each fuel tank must have a(n) _____ % expansion space.

5.  If the design landing weight of an aircraft is less than the permitted takeoff weight a fuel _____ system must be provided.

6.  The measure of the tendency of a liquid to vaporize under given conditions is known as _____.

7.  Detonation may also be known as engine _____.

8.  When considering aviation fuel performance numbers, the first number indicates the _____ (rich or lean)-mixture rating.

9.  The performance number ratings of the three grades of aviation gasoline in current use are:

    a.  _____

    b.  _____

    c.  _____

10. What color dye is used to identify each of these aviation gasolines:

    a.  Avgas 80    _____

    b.  Avgas 100 _____

    c.  100LL        _____


11. The two types of turbine fuel currently being used are:

    a.  _____

    b.  _____


12. Water occurs in aviation fuel in these two ways:

    a.  _____

    b.  _____


13. The most effective method to limit microbial growths in turbine fuel is to eliminate
    _____ from the fuel.


14. Low wing aircraft _____ (can or cannot) use gravity to feed fuel to the
    carburetor.


15. The electric boost pump is used for these three operations:

    a.  _____

    b.  _____

    c.  _____


16. Three types of aircraft fuel tanks are:

    a.  _____

    b.  _____

    c.  _____


17. Welded aluminum fuel tanks are usually made of _____ or
    _____ aluminum alloy.

18. Welded or riveted fuel tanks may be sealed by pouring a liquid sealant into the tank and allowing it to cover the entire inside of the tank. The surplus is poured out, and that which remains is allowed to cure. This method of sealing is called _____ the tank.

19. Fuel tanks that consist of part of the aircraft structure which is sealed off to hold the fuel are called _____ tanks.

20. A fuel bay that holds a bladder tank must have all of the edges of the metal and all of the rivets and screw heads covered with _____ tape.

21. If a bladder-type fuel tank is to be left empty for an extended period of time, the inside of the tank should be coated with a film of _____.

22. A fuel tank cap that has a "gooseneck" tube over the vent must be installed so the opening in the tube faces _____ (forward or rearward).

23. Most of the rigid fuel lines used in aircraft are made of _____ aluminum alloy.

24. A flareless fitting in a fuel line must be tightened no more than one-_____ _____ to one-_____ of a turn after the fitting is finger tight.

25. The _____ line of a flexible hose will indicate if the hose has been installed with a twist.

26. If an electrical wire bundle and a fuel line are run parallel through a compartment of an aircraft, the electrical wire bundle should be installed _____ (above or below) the fuel line.

27. "All runs of rigid tubing installed between fittings in an aircraft fuel system should have at least one bend in them." This statement is _____ (true or false).

28. A device that gives a fuel valve a positive feel when it is in the full ON or full OFF position is called a(n) _____.

29. Hand operated fuel valves found on small and medium sized aircraft will likely be one of these types:

    a. _____

    b. _____

30. Two types of motor-driven fuel valves are:

    a. _____

    b. _____

31. A wobble pump is a _____ (single or double) -acting pump.

32. The most popular type of auxiliary fuel pump is the _____ boost pump.

33. A centrifugal boost pump is used in its _____ (low or high) -speed position for supplying the fuel required to start the engine.

34. A fuel ejector system uses the _____ principle to supply additional fuel to the collector can.

35. Pulsating-type electric fuel pumps are installed _____ (in series or parallel) with the engine-driven fuel pump.

36. A vane-type pump is a _____ (constant or variable) displacement pump.

37. If a vane-type pump is installed in an airplane as an auxiliary fuel pump, it is installed in _____ (series or parallel) with the engine-driven fuel pump.

38. Two auxiliary valves that are built into a vane-type fuel pump are:

    a. _____

    b. _____

39. If a fuel filter in a jet aircraft clogs because of dirt in the fuel, the flow of fuel to the engine _____ (will or will not) be stopped.

40. When it is necessary to known the mass of the fuel in the tanks, a(n)
    _____ -type fuel quantity indicating system is used.

41. Reciprocating engines equipped with fuel injection systems may use a fuel
    flowmeter that is actually a fuel _____ gauge.

42. Before any fuel tank is welded, it must be purged of all explosive fumes using
    _____ or some chemical compound.

43. Welded fuel tanks are normally tested with _____psi of air pressure inside the
    tank.

44. Terneplate is used to manufacture some fuel tanks. It is sheet steel coated with
    _____ and _____.

45. Before repairing an integral fuel tank, it should be purged of all fuel vapors with
    either of these gases:
    a. _____
    b. _____

46. When testing a repaired integral fuel tank, it should have no more than _____
    psi air pressure put into the tank.

47. When classifying fuel leaks, the size of the surface area that a fuel leak moistens in
    a(n) _____ (how many) minute period is used.

48. The common forms of aviation fuel contaminants are:
    a. _____
    b. _____
    c. _____
    d. _____

49. The principle effects of micro organisms in turbine fuel are:
    a. _____
    b. _____
    c. _____

50. Aircraft _____ (should or should not) be serviced with fuel in a hangar or other enclosed areas.

51. The bleeding off of electrical charges on aircraft _____ (is or is not) an instantaneous act.

52. When connecting bonding cables between a refueler and an aircraft, you should connect the fueler to _____ (ground or aircraft) first.

# Chapter XVI
# Aircraft Cabin Atmospheric Control Systems

1. The two most abundant elements in our atmosphere are:

    a. _____

    b. _____

2. Express the standard sea level conditions of our atmosphere:

    a.  Sea level pressure      =      _____ pounds per square inch

    b.  Sea level pressure      =      _____ inches of mercury

    c.  Sea level temperature   =      _____ degrees Fahrenheit

    d.  Sea level temperature   =      _____ degrees Celsius

3. Above 36,000 feet, the air temperature stabilizes at _____ degrees Fahrenheit.

4. Some form of supplemental oxygen is normally recommended for flight above:

    a.  _____ feet in the daytime

    b.  _____ feet at night

5. When a person is hyperventilated, they suffer from a lack of _____ in the blood.

6. _____ is the product of incomplete combustion, and is found in varying amounts in the smoke and fumes from burning aviation fuels.

7. Pure oxygen _____ (will or will not) burn.

8. Aircraft oxygen systems should never be serviced with any oxygen that is not labeled _____ Oxygen.

9. Almost all military aircraft now carry oxygen in its _____ (solid, liquid, or gaseous) state.

10. Mechanically-separated oxygen systems produce oxygen by drawing air through a patented material called a(n) _____.

11. High pressure oxygen cylinders should be painted _____ (green or yellow).

12. What DOT approval codes should appear on aircraft oxygen cylinders?

    a. _____

    b. _____

13. Oxygen cylinder must be hydrostatically tested to _____ (what amount) of their working pressure.

14. The two types of continuous flow oxygen regulators are:

    a. _____

    b. _____

15. The oxygen regulator used by the flight crews of most commercial jet airliners are of the _____ (continuous flow or diluter-demand) -type.

16. A rebreather bag type mask is used with a _____ (continuous flow or diluter-demand) -type regulator.

17. The oxygen masks that automatically drop from the overhead compartment of a jet transport aircraft are of the _____ (continuous flow or demand) -type.

18. Most of the rigid plumbing lines that carry high-pressure oxygen are made of _____.

19. Liquid oxygen systems are referred to using the initials _____.

20. The pressure of the oxygen delivered to the regulator by a liquid oxygen converter is usually _____ psi.

21. _____ (what chemical) mixed with appropriate binders and a fuel is used to make oxygen candles.

22. Oxygen candles have a _____ (short or long) shelf life.

23. When servicing an aircraft oxygen system from a service cart having several cylinders, always start with the cylinder having the _____ (lowest or highest) pressure.

24. When a nitrogen cylinder is installed on the oxygen servicing cart it will face the _____ (same or opposite) direction of the oxygen cylinders.

25. An oxygen system is considered to be empty when the pressure gets down to _____ to _____ psi.

26. With an outside air temperature of 30°F. an oxygen system must be filled to _____ psi to obtain a stabilized pressure of 1,800 psi at 70°F.

27. During oxygen servicing there should be no smoking, open flame, or items which may cause sparks within _____ feet or more.

28. Aircraft cabin pressurization is controlled by varying the amount of air that is _____ (taken in or allowed to leak out).

29. The two modes of operation for cabin pressurization are:

    a. _____

    b. _____

30. Smaller reciprocating engine aircraft may obtain air for cabin pressurization from:

    a. _____

    b. _____

31. Most pressurization systems have these three cockpit indicators:

    a. _____

    b. _____

    c. _____

32. The _____ (outflow or safety) valve is controlled by the cabin altitude controller.

33. The most common type of heater for small single-engine aircraft is the _____ _____ heater.

34. Light- and medium-twin engine aircraft are often heated with _____ _____ heaters.

35. The thermostat of a combustion heater controls the _____ (fuel valve or ignition).

36. The ventilating air pressure inside the combustion heater is _____ (higher or lower) than the pressure of the combustion air.

37. When the temperature is reached for which the duct limit switch of a combustion heater is set, the _____ (fuel or ignition) to the heater is shut off.

38. Which of these controls for a combustion heater is not normally accessible in flight? _____ (A, B, or C)

    a.  Heater master switch

    b.  Cabin thermostat

    c.  Overheat switch

39. Two types of air conditioning system for lowering the temperature of the air in an aircraft cabin are:

    a. _____

    b. _____

40. Two methods of removing heat from the air in an air cycle air conditioning system are:

    a. _____

    b. _____


41. The water separator in an air cycle air conditioning system removes the water from the air _____ (before or after) it passes through the air cycle machine.


42. Heat that is added to a liquid that raises its temperature is called _____ _____ heat.


43. Heat that is added to a liquid that causes a change in its state without changing its temperature is called _____ heat.

44. Identify each of the numbered components in this vapor cycle air conditioning system.

    a.    1 is the _____

    b.    2 is the _____

    c.    3 is the _____

    d.    4 is the _____

    e.    5 is the _____

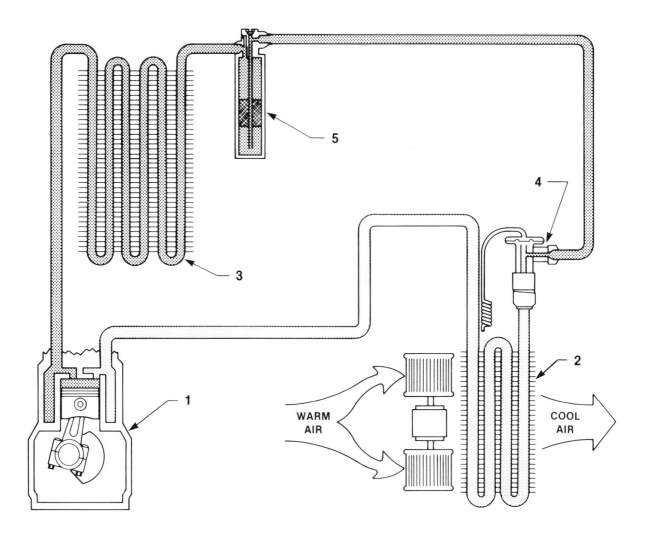

45. The two components in a vapor cycle air conditioning system that divide the system into the high and low side are the:

    a. _____

    b. _____

46. The component in a vapor cycle air conditioning system that acts as the reservoir for the refrigerant is the _____.

47. The component in a vapor cycle air conditioning system in which heat from the cabin is absorbed into the refrigerant is the _____.

48. The component in a vapor cycle air conditioning system in which the heat in the refrigerant is given up to the outside air is the _____.

49. The component in a vapor cycle air conditioning system in which the desiccant is held to remove water from the refrigerant is the _____.

50. The refrigerant most generally used in aircraft air-conditioning systems is _____ (R-12 or R-22).

51. The refrigerant used in an air conditioning system is _____ (flammable or non-flammable).

52. R-12 vapors are _____ (toxic or non-toxic).

53. R-12 vapors which have passed through a flame are _____ (toxic or non-toxic).

54. The thermal expansion valve of a vapor cycle air conditioning system is sometimes referred to as the _____ valve.

55. Two types of leak detectors that are suitable for detecting a refrigerant leak in an aircraft air conditioning system are:

    a. _____

    b. _____

56. If bubbles are visible in the _____ there is not enough refrigerant in the system.

57. Refrigerant should normally be put into the low side of an air conditioning system in its _____ (liquid or vapor) state.

58. Any time a vapor cycle air conditioning system has been opened, it must be _____ before it is recharged.

# CHAPTER I

1. Aerodynamic
2. Attack
3. Low
4. Behind
5. Spar
6. Compression
7. Tensile
8. Truss
9. Former rib
10. Drag
11. Cantilever
12. Chemical, electrochemical
13. Stiffness
14. Integral
15. a. Vertical or yaw
    b. Longitudinal or roll
    c. Lateral or pitch
16. Lateral
17. Downward
18. Up
19. Longitudinal
20. Up
21. Up
22. Frise
23. Vertical
24. Adverse
25. Rudder
26. Right
27. Ahead of
28. Corrugating
29. Inboard
30. Spoiler
31. Empennage
32. Vertical
33. Rudder
34. Stabilizer
35. Elevators
36. Dorsal
37. Stabilator
38. Anti-servo
39. Ruddervators
40. Drag
41. Increase
42. Slotted
43. Fowler
44. Camber
45. Kruger
46. Aileron
47. a. Aerodynamic factors
    b. Mechanical actuation
48. At the root
49. Vortex generators
50. a. Ailerons
    b. Elevators
    c. Rudders
51. Fixed
52. Down
53. Opposite
54. Opposite
55. Same
56. High
57. Large
58. Pratt
59. Warren
60. Monocoque
61. Semi-monocoque
62. Fail-safe
63. Conventional
64. Tricycle
65. Parasite
66. Pressure
67. Fins
68. Cowl flaps
69. Open
70. a. Under the wings
    b. On the aft fuselage
71. Composite

# CHAPTER II

## PART ONE

1. 78,21
2. a. 14.69
   b. 29.92
   c. 1013.2
3. Altimeter
4. a. Celsius
   b. Kelvin
   c. Fahrenheit
   d. Rankine
5. a. 37.8
   b. 422.6
   c. 5007
   d. 240
6. a. 15
   b. 59
7. Relative humidity
8. Less
9. Slugs
10. 0.002378 slugs
11. Density
12. a. Kinetic
    b. Potential
13. Potential
14. Kinetic
15. Decrease
16. Noncompressible
17. 1/3
18. Stagnation
19. a. Surface area of the airfoil
    b. Lift coefficient of the airfoil
    c. Dynamic pressure of the air
20. a. Shape of the airfoil
    b. Angle of attack
21. q
22. $q = pV^2$
23. Boundry
24. a. Induced
    b. Parasite
25. Decreases
26. Increases
27. Symmetrical
28. a. Longitudinal
    b. Lateral
    c. Vertical
29. a. Static
    b. Dynamic
30. a. lateral
    b. Longitudinal
    c. Vertical
31. Right
32. a. Longitudinal
    b. Lateral
    c. Vertical
33. Rearward
34. Up
35. Balance
36. Frise
37. Springs
38. Opposite
39. Anti-servo
40. The same
41. Down
42. Fowler
43. Slot
44. Root
45. Wing fence
46. a. Ailerons
    b. Spoilers
47. Servo tabs
48. 2
49. Yaw damper
50. a. Kruger flaps
    b. Retractable slats
51. Compressible
52. 661.7
53. Mach number
54. a. Subsonic
    b. Transonic
    c. Supersonic
55. Decreases
56. Decreases
57. Decreases
58. 50%
59. Top
60. Critical Mach number
61. Normal
62. Aerodynamic heating

## PART TWO

1. Aerodynamic
2. a. Tail rotor
   b. Two main rotors
3. Feather
4. a. Gravity
   b. Centrifugal force
   c. Lift
   d. Gyroscopic forces
5. Symmetrical
7. a. Rigidity
   b. Precession
6. Precession
8. Coriolis effect
9. Drag
10. Underslung
11. Tail rotor
12. Inside ground effect
13. Outside ground effect
14. 1/2
15. Forward flight
16. Flapping
17. Advancing
18. Retreating
19. High-
20. Translational
21. Autorotation
22. Downward
23. Upward
24. Collective
25. Cyclic
26. Synchronized elevators
27. Irreversible
28. Pedals
29. Unatable
30. a. Stabilizer bar
    b. Offset flapping hinge
    c. Stability augmentation system
31. Beat
32. Low
33. Lateral
34. Vertical
35. a. Chordwise
    b. Spanwise
36. a. Marking stick
    b. Tracking flag
    c. Strobe light
37. On the ground
38. In flight
39. Fan
40. Collective
41. a. Direct shaft
    b. Free turbine
42. Clutch
43. Freewheeling

## PART THREE

1. Incidence
2. Increased
3. Type Certificate Data Sheet
4. Droop
5. a. 1 ∞ 7; 1 ∞ 19
   b. 7 ∞ 7
   c. 7 ∞ 19
6. a. 75
   b. 100
   c. 100
7. 60%
8. a. Wear
   b. Corrosion
9. May not
10. a. 100
    b. 50
    c. 74
    d. 23
11. 3
12. 4
13. Is not
14. Positive
15. Decalage
16. Landing
17. Lower

## PART FOUR

1. Design
2. Are
3. 91
4. March 31, 1992
5. Is
6. FAR Part 43, Appendix D
6. Is not
8. 24
9. 24
10. E
11. 100 hours
12. Are
13. Are not

# CHAPTER III

1. Aluminum
2. Nitrate
3. Is
4. a. Manufacturer's service manual
   b. AC 43.13-1A
   c. Supplemental Type Certificate
5. a. Restoration of old aircraft
   b. Amateur built aircraft
   c. Commercial recovering
6. 80
7. 56
8. Glider
9. Reinforcing
10. Surface

11. Second
12. Retarder
13. Thinned
14. 5
15. Plasticizers
16. Will not
17. Spar varnish
18. a. Blanket
    b. Envelope
19. Baseball
20. Water
21. Loosen
22. Animal
23. 250
24. 1 3/4, 2 1/2
25. Beside
26. Seine
27. Splice
28. Grounded
29. 200
30. Cutting
31. Much
32. Clear
33. Glossy
34. Polyester
35. a. 3.7
    b. 2.8
36. Heat
37. Non-taughtening
38. Butyrate
39. Sprayed
40. Baseball
41. 4, 8 to 10
42. Sew

# CHAPTER IV

1. Plasticizers
2. Flexibility
3. Methyl Ethyl Ketone (MEK)
4. Solvents, plasticizers
5. Heavy
6. 20
7. Too much
8. Water
9. Retarder
10. Warming
11. a. Dope exposed to too much
    heat
    b. Excessive atomizing air
    pressure on the spray gun
12. a. Too much dope applied
    b. Spray gun held too close
    c. Dope was improperly
    thinned
13. a. Improper spray technique
    b. Use of thinner that
    evaporates too fast
14. Fisheyes
15. Roping
16. a. Cracking
    b. Difficult to remove for
    repairs
17. Flexitive
18. Sandpaper
19. Enamel
20. a. Aluminum foil
    b. Polyethylene sheeting
21. Thick
22. Hot water, steam
23. Plastic
24. Polyurethane
25. Conversion

26. Acrylic
27. Low
28. B
29. Polyurethane
30. Wash
31. Induction
32. Viscosity cup
33. 5
34. Pot
35. a. Primer
    b. Acid diluent
    c. Thinner
36. 0.3
37. 8
38. Moisture
39. a. White reflective basecoat
    b. Colored pigments
    c. Clear topcoat
40. Polyurethane
41. Toluol
42. Will not
43. Linseed
44. Heavier
45. Daily
46. a. Air spray
    b. Airless
    c. Electrostatic
47. Will not
48. Air, electric
49. Perpendicular
50. a. 4
    b. 5
    c. 2
    d. 1
51. Much
52. True
53. 2/3
54. Ahead
55. Acetone, MEK
56. 45
57. N
58. 58
59. Bad
60. Water
61. Wet

# CHAPTER V

1. Aluminum
2. a. Monocoque
    b. Semi-monocoque
3. a. Strength
    b. Stiffness
    c. Shape
4. a. Tension
    b. Compression
    c. Bending
    d. Torsion
    e. Shear
5. Compressive
6. Tensile
7. Tension, compression
8. Shear
9. Stop-drill
10. Are
11. a. Copper
    b. Copper
    c. Magnesium
    d. Zinc
12. 1100
13. 2024
14. Solution

15. Precipitation
16. Softened
17. a. F
    b. O
    c. T4
    d. T3
    e. T6
    f. H12
    g. H18
18. a. Cladding
    b. Cover with an oxide film
    c. Cover with an organic
    coating
19. a. Electrode potential
    difference
    b. Conductive path between
    the pieces
    c. An electrolyte present
20. Will not
21. a. Electrostatically
    b. Chemically
22. Stiffness
23. Lighter
24. a. It corrodes easily
    b. It is difficult to work with
    c. It is highly flammable
25. Should not
26. Are
27. Transfer
28. a. Green
    b. Red
    c. Yellow
29. a. Tip
    b. Body
    c. Shank
30. 30
31. Cornice
32. a. Silver
    b. Copper
    c. Black
    d. Brass
33. Hole finder
34. a. 100° countersunk
    b. Universal
    c. Round
    d. Flat
35. a. 1100
    b. 2117
    c. 2017
    d. 2024
    e. 5056
36. a. 1100
    b. 5056
    c. 2117
    d. 2017
    e. 2024
    f. 7050
37. B
38. 1/16
39. 1/32
40. B
41. 186
42. 858
43. a. 245
    b. 328
    c. 409
44. 2
45. 3, 10 to 12
46. 3/4
47. a. #40
    b. #30
    c. #21
    d. #11

48. 100
49. Dimpled
50. Coin
51. Hot
52. One-shot
53. 3,4
54. a. 1/2
    b. 1 1/2
55. Inside
56. a. 3
    b. 2
    c. 1
57. Scross
58. a. 1/16
    b. 3/32
    c. 5/16
59. Mold
60. Setback
61. Bend radius, thickness
62. K
63. K(BR + MT)
64. a. 0.290
    b. 0.120
    c. 0.70018
    d. 0.625
65. Bend allowance
66. a. 0.421
    b. 0.218
    c. 0.328
    d. 0.214
    e. 0.642
67. Shrunk
68. May
69. Flanged
70. Stop-drilled
71. Burnishing
72. 337
73. An octagon does not cause an
    abrupt change in the cross
    sectional area as stresses
    enter and leave the repair.
74. Aircraft bolts and self-locking
    nuts
75. 1/8
76. Inboard
77. Slip roll former
78. Wet

# CHAPTER VI

1. Rot-inhibiting
2. Spar varnish
3. Mahogany
4. Lowest
5. Rotted
6. Sitka spruce
7. a. Plastic resin
    b. Resourcinol
8. Should not
9. a. Pot life
    b. Open assembly time
    c. Closed assembly time
10. May not
11. a. 4 1/2"
    b. 9"
    c. 1 1/2"
12. 1
13. 1/10
14. 5
15. 12
16. a. Thermosetting

b. Thermoplastic
17. Thermosetting
18. a. Natural
    b. Cellulose
    c. Protein
    d. Synthetic
19. Linen, phenolic
20. a. E-glass
    b. S-glass
21. Aramid
22. Compressive
23. Ceramic
24. Parallel
25. Warp
26. Selvage
27. Bias
28. Unidirectional
29. Mats
30. More
31. a. Fiberglass lay-up
    b. Compressive molding
    c. Vacuum bagging
    d. Filament winding
    e. Wet lay-up
32. a. Aluminum wires may be
       woven into the top layer of
       composite fabric.
    b. A fine aluminum screen
       may be laminated under the
       top layer of fabric.
    c. A thin aluminum foil sheet
       may be bonded to the outer
       layer composite
    d. Aluminum may be sprayed
       onto the component.
    e. A piece of metal may be
       bonded to the composite.
33. Inhibitor
34. Heat
35. Catalyzed resin
36. Thick
37. More
38. 65, 85
39. Curing
40. Microballoons
41. Resin
42. Chopped fibers, flox
43. a. Aluminum
    b. Kevlar
    c. Carbon
    d. Fiberglass
    e. Paper
    f. Nomex
    g. Steel
44. Epoxy
45. Urethane
46. a. Negligible
    b. Repairable
    c. Non-repairable
47. Coin tap
48. Thermography
49. May
50. Cannot
51. Routers
52. MEK
53. Delamination
54. Will
55. Potting
56. Balsa wood
57. Sealer
58. a. Structural strength
    b. Aerodynamic smoothness

c. Electrical transparency
59. Temporary
60. Unairworthy
61. Sanding
62. Step
63. Oil, grease
64. Ribbon
65. Warp compass
66. Ohmmeter
67. a. Remove excess resin
    b. Remove trapped air
       between layers
    c. Maintain the contour of the
       surface
    d. Prevent shifting during the
       curing
    e. Compact the fiber layers
       together.
68. Vacuum bagging
69. Release
70. Bleeders
71. a. Room temperature curing
    b. Heat curing
72. Step
73. Is not
74. Heat, pressure
75. Blankets
76. Aramid
77. Should not
78. Backed
79. 135, high
80. Aramid
81. Should not
82. Should not
83. Carbide
84. Discarded
85. Respirators
86. a. Acrylic
    b. Acetate
87. Will not
88. C
89. Crazed
90. Unsatisfactory
91. 150
92. Ethylene dichloride
93. Does
94. Heat treating
95. Mild soap

## CHAPTER VII

1. Fusion
2. Brazing
3. 800
4. a. Gas
   b. Electric arc
   c. Electric resistance
5. a. Oxygen
   b. Acetylene
6. Stick
7. Gas Metal
8. Gas Tungsten, TIG
9. 10,000
10. MIG
11. DC
12. Tungsten
13. a. Spot welding
    b. Seam welding
14. a. Amount of current
    b. Pressure applied
    c. Dwell time
15. 15

16. 4,8
17. Acetone
18. Weight
19. 1/4, 1/2
20. Cleaner
21. Petroleum
22. 6,300
23. Right
24. Left
25. Open
26. Red, left
27. Green, right
28. Acetylene
29. Balanced-pressure
30. Tip
31. Twist drill
32. Smaller
33. Lighter
34. Blue
35. May not
36. Copper
37. Thickness
38. Neutral
39. Lap
40. a. Uniform width
    b. Good penetration
    c. Good reinforcement
    d. Uniform ripples
41. 100
42. Hydrogen
43. Does not
44. Carburizing
45. Neutral
46. Capillary
47. B
48. Tin, lead
49. Silver
50. Inert gas
51. Straight
52. Pointed
53. AC
54. Rounded
55. DC
56. 4130
57. 43.13-1A
58. Scarf, fishmouth

## CHAPTER VIII

1. Spoiler
2. Does not
3. Anti-icing
4. De-icing
5. a. Hot air
   b. Electric current
   c. Chemical
6. Bleed
7. Bleed air
8. Electrical heaters
9. Alternate
10. Electrically
11. Heated
12. a. Carburetor
    b. Propeller
    c. Windshield
13. Isopropyl
14. Vacuum
15. Suction
16. Bleed
17. a. Adhesives
    b. Screws and rivnuts
18. Soap, water

19. Electrothermal
20. Slip rings, brush blocks
21. On a sequenced cycle
22. Ethylene glycol, isopropyl
    alcohol
23. a. Windshield wipers
    b. Chemical rain repellant
    c. High velocity air blast
24. a. Electric
    b. Hydraulic
25. Heavy
26. Bleed

## CHAPTER IX

1. a. Light weight
   b. Ease of installation
   c. Simple inspection
   d. Minimum maintenance
2. 100
3. Incompressible
4. Height of the column
5. Pressure
6. $F = P \infty A$
7. Volume = Area X Distance
8. a. 1,000
   b. 1/2
   c. 100
   d. 500
9. a. Up
   b. 1,000
10. a. 5,000
    b. 5
    c. 4,000
    d. 72
    e. 40
    f. 33.3
    g. 0.1
    h. 0.4
    i. 2,500
    j. 4
    k. 100
    l. 1.1
    m. 0.002
    n. 500,000
    o. 6
    p. 50,000
11. a. Able to flow through all of
       the lines with a minimum of
       opposition.
    b. Be incompressible.
    c. It must have good lubricat-
       ing properties.
    d. It must inhibit corrosion
    e. It must not foam in
       operation.
12. Viscosity
13. Flash point
14. a. Vegetable base
    b. Mineral base
    c. Synthetic
15. Castor, alcohol
16. Blue
17. Mineral base
18. Mineral base
19. Fire
20. Light purple
21. a. Alcohol
    b. Naptha, Vasol, Stoddard
       solvent
    c. Trichlorethylene
22. a. Natural rubber
    b. Neoprene

c. Butyl
23. Soap, water
24. a. Reservoir
    b. Pump
    c. Actuators
25. Power pack
26. a. Integral
    b. In-line
27. Unpressurized
28. a. Aspirator
    b. Compressor bleed air
    c. System hydraulic pressure
29. 0.000039
30. Return
31. Can
32. Constant
33. Constant
34. Unloading
35. Does not
36. Open
37. Check
38. Orifice check
39. Sequence
40. Priority
41. Hydraulic fuse
42. a. Rate of flow
    b. Volume of flow
43. Relief
44. Pressure reducer
45. a. Piston
    b. Bladder
    c. Diaphragm
46. Air, nitrogen
47. H
48. Linear
49. Motor
50. One-way
51. Two-way
52. Larger
53. Away from
54. Cure
55. Would
56. High
57. Bleed
58. Vane
59. Dessicant, chemical dryer
60. Shuttle
61. Shuttle

## CHAPTER X

1. Parasite
2. Conventional
3. Tricycle
4. Wheel pants
5. Bungee
6. Air-oil, oleo
7. Oil
8. Air
9. Strut extension
10. Toed-in
11. Negative
12. Safety
13. Retarded, up
14. Shimmy damper
15. Magnesium, aluminum
16. Bead seat
17. Fusible plugs
18. Should not
19. Intergranular
20. Heat
21. Is not

22. Magnetic particle
23. Any
24. Aircraft
25. Debooster
26. Pneumatic
27. Linings
28. 0.100
29. Spongy
30. Bleeding
31. Wheel cylinder
32. a. Flush the system
    b. Replace all seals
33. Fluorescent penetrant
34. Is not
35. Overheating
36. Dragging
37. a. Wheel-speed sensors
    b. Control box
    c. Control valves
38. a. Generate electrical signals
       usable by the control valve.
    b. Control brake pressure to
       prevent a skid.
    c. Prevent brake pressure
       from being applied prior to
       touchdown
39. Is
40. III
41. Does not
42. Tubeless
43. More
44. Rib
45. Nose
46. Proper inflation
47. a. 1
    b. 3
    c. 2
48. Under-inflated
49. Airframe
50. Cold
51. Is
52. O-ring
53. Are
54. Cannot
55. Is not
56. Is
57. Vertically
58. a. A hole in the tube
    b. A defective valve
59. Valve
60. Light
61. Anti-sieze
62. Heavy
63. a. On brackets under the head
       of the wheel bolts
    b. With cotter pins through
       holes in the wheel rim
    c. With double surface tape
64. Taxiing
65. Hydroplaning

## CHAPTER XI

1. a. Heat
   b. Fuel
   c. Oxygen
2. a. C
   b. B
   c. D
   d. A
3. Spot
4. Pre-set temperature

5. Parallel
6. Edison
7. Rate-of-temperature-rise
8. a. Sensitive
   b. Slave
9. Series
10. Reference
11. Continuous loop
12. Kidde
13. Sensor-responder
14. a. Lindberg
    b. Systron-Donner
15. D
16. a. Light refraction
    b. Ionization
    c. Solid-state
17. Ionization
18. Halon
19. Dry powder
20. a. Water
    b. Carbon dioxide
    c. Dry chemical
    d. Halogenated hydrocarbons
21. Carbon dioxide
22. a. Conventional
    b. High-rate-of-discharge
       (HRD)
23. Yellow
24. May not
25. Weighing
26. 5

## CHAPTER XII

1. Amps
2. Resistance, ohms
3. Volts
4. $E = I \times R$
5. Watts
6. 746
7.

| | | | | | | |
|---|---|---|---|---|---|---|
| **WATTS** | 18 | 150 | 5,600 | 144 | 120 | 360 |
| **OHMS** | 2 | 105 | 14 | 4 | 7.5 | 40 |
| **AMPS** | 3 | 10 | 20 | 6 | 4 | 3 |
| **VOLTS** | 6 | 15 | .280 | 24 | 30 | 120 |

8. a. Chemical action
   b. Heat
   c. Pressure
   d. Light
   e. Magnetic
9. A
10. a. Flux
    b. Electron flow
    c. Conductor movement
11. a. Flux

b. Electron flow
   c. Conductor movement
12. 400
13. a. Resistance
    b. Inductive reactance
    c. Capacitive reactance
14. Impedance, Z
15. Are
16. Lead
17. Lag
18. a. Frequency of the AC
    b. Inductance of the coil
19. a. Frequency of the AC
    b. Capacity
20. Increases
21. Decreases
22. Power factor
23. $P = E \times I$
24. $P = E \times I \times$ Power factor
25. Watts
26. Volt-amps or KVA
27. 2.1
28. 1.275, 1.300
29. 1.150
30. Low
31. Is not
32. Temperature
33. Brushes, commutator
34. Reverse current-cutout
35. Field
36. a. Voltage regulator
    b. Current limiter
    c. Reverse current-cutout
       relay
37. a. Closed
    b. Closed
    c. Open
38. Solid-state
39. Is not
40. 400, 3
41. Brushless
42. Constant speed drive
43. Voltage spikes
44. A
45. a. Supply all electrical loads
    b. Keep the battery charged
46. Relay
47. A
48. Contactor
49. a. False
    b. False
    c. True
    d. True
50. 26
51. Coil
52. Voltage
53. a. Split-bus
    b. Parallel
54. Parallel
55. a. Transformer
    b. Circuit breaker
    c. DPST switch
    d. Relay (N.O.)
    e. Fuse
    f. Capacitor
    g. Iron-core inductor
    h. Shielded wire
    i. Battery
    j. Wire disconnect
    k. Silicon controlled rectifier
    l. Zener diode
    m. Thermocouple

n. PNP transistor
o. Variable capacitor
56. 600
57. 6
58. American Wire
59. Shielding
60. Coaxial
61. a. 2.0
    b. 0.5
    c. 8.0
    d. 7.0
62. a. 4
    b. 2
    c. 10
    d. 2/0
63. Above
64. 25
65. 0.003
66. a. Red
    b. Yellow
    c. Blue
67. 4
68. BNC
69. AC 43.13-1A
70. 2
71. Relay
72. Slo-blow
73. a. Push-to-reset
    b. Push-pull
    c. Toggle
74. Trip free
75. a. Permanent magnets
    b. Electromagnets
76. a. Series
    b. Shunt
    c. Compound
77. High
78. Series
79. a. The design of the motor
    b. The frequency of the AC applied
80. a. Red
    b. Green
    c. White
81. a. Rotating beacon
    b. Flashing or strobe-light
82. 5

# CHAPTER XIII

1. Absolute
2. a. Pitot
   b. Static
3. Absolute
4. Engine Pressure Ratio
5. Altimeter
6. Indicated
7. Pressure
8. 29.92
9. Mean sea level
10. 32,000
11. Is not
12. Temperature
13. 24
14. Indicated
15. Static
16. a. Airspeed indicator
    b. Altimeter
    c. Vertical speed indicator
17. 43
18. 100
19. Mach

20. 0.95
21. Vertical speed
22. Wheatstone bridge, ratiometer
23. Thermocouple
24. Is not
25. Accelerometer
26. a. Autosyn
    b. Magnesyn
27. Magnesyn
28. Synchronoscope
29. 1/2
30. Electronic
31. a. Rigidity
    b. Precession
32. a. Pitch
    b. Roll
33. Rigidity
34. Precession
35. Inclinometer
36. Vertical
37. The turn coordinator's gimble axis is tilted about 30° so it will precess when the aircraft rolls as well as when it yaws.
38. North-South
39. Variation
40. Isogonic
41. Agonic
42. Deviation
43. Deviation
44. Should
45. In-flight
46. Carbon
47. Resistance
48. The fuel and air
49. 1
50. 2
51. psi
52. Most
53. Stall
54. a. Error sensing
    b. Correction
    c. Follow-up
    d. Command
55. a. Attitude gyros
    b. Rate gyros
56. Yaw damper
57. a. 3
    b. 1
    c. 4
    d. 2
58. a. 4
    b. 1
    c. 3
    d. 5
    e. 2
59. a. Electronic Flight Instrument System
    b. Cathode Ray Tube
    c. Electonic Attitude Director Indicator
    d. Aeronautical Radio Incorporated
    e. Electronic Horizontal Situation Indicator
60. Zulu

# CHAPTER XIV

1. Analog
2. Digital
3. Check
4. Forward

5. a. Emitter
   b. Collector
   c. Base
6. a. Emitter
   b. Collector
   c. Base
7. PNP
8. Forward
9. Positive
10. High
11. a. Light emitting diode
    b. Photo diode
12. a. Rectifier
    b. Amplifier
    c. Oscillator
13. Rectifier
14. Oscillator
15. Amplifier
16. a. Handle information
    b. Perform work
17. Logic gates
18. a.

| A | B | C |
|---|---|---|
| 0 | 0 | 0 |
| 1 | 0 | 0 |
| 0 | 1 | 0 |
| 1 | 1 | 1 |

b.

| A | B | C |
|---|---|---|
| 0 | 0 | 0 |
| 1 | 0 | 1 |
| 0 | 1 | 1 |
| 1 | 1 | 1 |

c.

| A | B | C |
|---|---|---|
| 0 | 0 | 1 |
| 1 | 0 | 1 |
| 0 | 1 | 1 |
| 1 | 1 | 0 |

d.

| A | B | C |
|---|---|---|
| 0 | 0 | 1 |
| 1 | 0 | 0 |
| 0 | 1 | 0 |
| 1 | 1 | 0 |

19. a. Electromagnetic
    b. Electrostatic
20. a. Amplitude modulation
    b. Frequency modulation
21. Intermediate frequency
22. VHF
23. Does
24. Shorter
25. Hertz
26. Quarter
27. Loop
28. VHF
29. Phase
30. a. Left-right
    b. To-from
31. LF/MF
32. a. Loop
    b. Sense
33. a. Localizer
    b. Glide slope
    c. Marker beacons
34. Localizer
35. Localizer

36. Voltage
37. 75 Mhz
38. Middle
39. VHF
40. Glide slope
41. Down
42. Is
43. VHF
44. 4096
45. C
46. a. VOR
    b. ADF
47. a. Green
    b. Yellow
    c. Red
48. Liquid water
49. X
50. X
51. Vortac
52. a. 121.5 Mhz
    b. 243 Mhz
53. a. On the outside of the ELT case
    b. In the aircraft maintenance records
54. LF
55. Do not
56. Null field dischargers
57. 0.003
58. Coaxial

# CHAPTER XV

1. Is not
2. 150
3. 125
4. 2
5. Jettisoning
6. Volatility
7. Knock
8. Lean
9. a. 80/87
   b. 100/130
   c. 100LL
10. a. Red
    b. Green
    c. Blue
11. a. Jet A and Jet A-1
    b. Jet B
12. a. Dissolved
    b. Free
13. Water
14. Cannot
15. a. Supply fuel for starting the engine
    b. Assure a positive pressure at the engine driven pump
    c. Assure a continuous flow which switching fuel tanks in flight
16. a. Welded or riveted
    b. Integral
    c. Bladder
17. 3003, 5052
18. Sloshing
19. Integral
20. Chafe resistant
21. Engine oil
22. Forward
23. 5052
24. sixth, third
25. Lay

26. Above
27. True
28. Detent
29. a. Cone
    b. Poppet
30. a. Drum
    b. Sliding gate
31. Double
32. Centrifugal
33. Low
34. Venturi
35. Parallel
36. Constant
37. Series
38. a. Relief
    b. Bypass
39. Will not
40. Capacitance
41. Pressure
42. Steam
43. 3.5
44. Lead, tin
45. a. Argon
    b. Carbon dioxide
46. 1/2
47. 30
48. a. Solid particles
    b. Surfactants
    c. Water
    d. Micro-organisms
49. a. The formation of sludge or
       slime.
    b. The emulsification of the
       fuel.
    c. The creation of corrosive
       compounds and offensive
       odors.

50. Should not
51. Is not
52. Ground

## CHAPTER XVI

1. a. Nitrogen
   b. Oxygen
2. a. 14.69
   b. 29.92
   c. 59
   d. 15
3. -69.7
4. a. 15,000
   b. 10,000
5. Carbon dioxide
6. Carbon monoxide
7. Will not
8. Aviators Breathing
9. Liquid
10. Molecular sieve
11. Green
12. a. DOT3AA1800
    b. DOT3HT1850
13. 5/3
14. a. Manual
    b. Automatic
15. Diluter demand
16. Continuous flow
17. Continuous flow
18. Stainless steel
19. LOX
20. 70
21. Sodium chlorate
22. Long

23. Lowest
24. Opposite
25. 50, 100
26. 1,725
27. 50
28. Allowed to leak out
29. a. Isobaric
    b. Constant pressure
       differential
30. a. Bleed air from the turbo-
       chargers
    b. Exhaust of the dry air pump
31. a. Cabin altitude
    b. Cabin rate of climb
    c. Pressure differential
32. Outflow
33. Exhaust shroud
34. Combustion
35. Fuel valve
36. Higher
37. Fuel
38. C
39. a. Air cycle
    b. Vapor cycle
40. a. Transfer heat into the air in
       the heat exchanger
    b. Using the heat energy to
       drive the compressor.
41. After
42. Sensible
43. Latent
44. a. Compressor
    b. Evaporator
    c. Condenser
    d. Thermal expansion valve
    e. Receiver-drier

45. a. Compressor
    b. Thermal expansion valve
46. Receiver-drier
47. Evaporator
48. Condenser
49. Receiver-drier
50. R-12
51. Non-flammable
52. Non-toxic
53. Toxic
54. TXV
55. a. Soap solution
    b. Electronic oscillator
56. Sight glass
57. Vapor
58. Evacuated